T0275684

Solid-Liquid Filtration

Solid-Liquid Filtration
Practical Guides in Chemical Engineering

Barry A. Perlmutter

AMSTERDAM • BOSTON • HEIDELBERG • LONDON
NEW YORK • OXFORD • PARIS • SAN DIEGO
SAN FRANCISCO • SINGAPORE • SYDNEY • TOKYO

Butterworth-Heinemann is an imprint of Elsevier

Butterworth-Heinemann is an imprint of Elsevier
The Boulevard, Langford Lane, Kidlington, Oxford OX5 1GB, UK
225 Wyman Street, Waltham, MA 02451, USA

Notices
Knowledge and best practice in this field are constantly changing. As new research and experience broaden our understanding, changes in research methods, professional practices, or medical treatment may become necessary.

Practitioners and researchers must always rely on their own experience and knowledge in evaluating and using any information, methods, compounds, or experiments described herein. In using such information or methods they should be mindful of their own safety and the safety of others, including parties for whom they have a professional responsibility.

To the fullest extent of the law, neither the Publisher nor the authors, contributors, or editors, assume any liability for any injury and/or damage to persons or property as a matter of products liability, negligence or otherwise, or from any use or operation of any methods, products, instructions, or ideas contained in the material herein.

ISBN: 978-0-12-803053-0

British Library Cataloguing-in-Publication Data
A catalogue record for this book is available from the British Library

Library of Congress Cataloging-in-Publication Data
A catalog record for this book is available from the Library of Congress

For Information on all Butterworth-Heinemann publications
visit our website at http://store.elsevier.com/

Working together
to grow libraries in
developing countries

www.elsevier.com • www.bookaid.org

CONTENTS

List of Figures vii
List of Tables ix
About the Authors xi
Acknowledgments xiii

Chapter 1 Introduction 1
Filtration Overview...2
Filtration Equipment Categories 2
Principles and Mechanisms...3
Filter Aids...8
Filter Media..9
Coagulants and Flocculants..15
Surface Charges 17
Filter Rating Systems 18

Chapter 2 Filtration Testing 21
Filtration Theory: Background and How to Use It in Practice 22
Slurry Characteristics..26
Particle Characteristics 28
Pressure Testing...31
Vacuum Testing..32

Chapter 3 Types of Filtration Systems...................... 35
Batch Systems..36
Continuous Systems...42

Chapter 4 Combination Filtration 47
Testing for Combination Filtration 47
Thickening..48
Polishing 52
Process Segmentation 53

Chapter 5 Filtration Selection..**57**
Specifications...60
Upstream and Downstream Equipment 67
Integration and Controls 68
Applications...71
Centrifugal Alternatives to Pressure and Vacuum Solid-Liquid
Filtration 72
Life Cycle Capital Equipment Costs...74

Chapter 6 Commissioning and Operation..............................**79**
Commissioning Plan..79
Preventative Maintenance..80
Troubleshooting..81
Interesting Process Challenges After the Fact..................................83

Chapter 7 Conclusion **89**

Appendix: Paint Filter Liquids Test.....................................**95**
Glossary of Important Filtration Terms...............................**99**
Suggested Further Reading Online **117**
References...**119**
Bibliography...**121**

LIST OF FIGURES

Figure 1.1	Selecting solid-liquid filtration equipment per Davies	4
Figure 1.2	Chase and Mayer's approach to separation	5
Figure 1.3	Inertial impaction	6
Figure 1.4	Diffusional interception	6
Figure 1.5	Direct interception	7
Figure 1.6	Particles striking a pore in direct interception	7
Figure 1.7	Plain, square weave	11
Figure 1.8	Twill weave	12
Figure 1.9	Plain, reverse Dutch	12
Figure 1.10	Double layer weave	13
Figure 1.11	Plain weave in metal media	14
Figure 1.12	Twilled weave in metal media	14
Figure 1.13	Plain Dutch weave in metal media	15
Figure 1.14	Twilled Dutch weave in metal media	16
Figure 2.1	Darcy equation	24
Figure 2.2	Simplified Darcy equation showing the relationship between filtration flow and time	24
Figure 2.3	Simplified Darcy equation showing the relationship between filtration flow and cake height	24
Figure 2.4	Ideal PSD with one main peak	30
Figure 2.5	Multiple (twin) peak and wide PSD	31
Figure 3.1	Three components of filter candles	39
Figure 3.2	Outer support tie rods provide annular space	40
Figure 3.3	Candle filter vessel construction	40
Figure 4.1	Testing for combination filtration	48
Figure 4.2	Concentrating candle filters followed by pressure plate batch filtration	49
Figure 4.3	Concentrating candle filters followed by continuous vacuum filtration	50
Figure 4.4	Concentrating candle filters followed by conventional filter press filtration	51
Figure 4.5	Chlor-alkali brine filtration with concentrating candle filters followed by conventional filter press filtration	51
Figure 4.6	Continuous vacuum filtration followed by candle filtration	52
Figure 4.7	Continuous rotary pressure filtration followed by candle filtration	53
Figure 4.8	Continuous vacuum filtration (two variations) followed by contained filter press	54
Figure 4.9	Continuous vacuum filtration (two variations)	54
Figure 5.1	Testing for garnet	58
Figure 5.2	Testing for wollastonite	59
Figure 5.3	Microscopy of glass spheres	60
Figure 6.1	Clarification application with candle filters	85
Figure A.1	Paint filter test apparatus	96
Figure A.2	Paint filter liquids test	97

LIST OF TABLES

Table 1.1	Typical beta ratio ranges	20
Table 5.1	Impact of particle size on process	58
Table 5.2	Mechanical properties checklist	61
Table 5.3	New filtration process questionnaire	61
Table 5.4	Existing filtration process questionnaire	64
Table 5.5	Sample application considerations	71
Table 6.1	Typical training outline	80
Table 6.2	Sample training quiz	81
Table 6.3	Typical mechanical preventative maintenance programs	82
Table 6.4	Typical process preventative maintenance programs	82
Table 6.5	Common filtration system symptoms	83
Table 7.1	Typical test data required	90
Table 7.2	Test setup	91
Table 7.3	Typical testing plan for pressure filtration using filter aid	92

Barry A. Perlmutter is currently President and Managing Director of BHS-Sonthofen Inc., a subsidiary of BHS-Sonthofen GmbH. BHS-Sonthofen is a manufacturer of filtration, washing, and drying technologies as well as mixing and recycling technologies. He has over 30 years of technical engineering and business marketing experience in the field of solid-liquid separation including filtration, centrifugation, and process drying. He has published and lectured extensively worldwide on the theory and applications for the chemical, pharmaceutical, and energy/environmental industries and has been responsible for introducing many European companies and technologies into the US marketplace. He began his career with the US Environmental Protection Agency and then Pall Corporation. He received a BS degree in Chemistry from University at Albany, State University of New York, an MS degree in Environmental Science Technology from the School of Engineering at Washington University, St. Louis, and an MBA from the University of Illinois, Chicago.

Detlef Steidl is currently the Director of Application Engineering for Filtration Technology at BHS-Sonthofen GmbH. He has over 25 years of experience with chemical manufacturing applications from the Max-Planck-Institute and Degussa Corporation in Mainz, Germany, and has been with BHS since 1988. He has an advanced engineering degree (Diploma in Process Technology) from Frankfurt University of Applied Sciences.

ACKNOWLEDGMENTS

The world of process filtration is very small and made up of dedicated individuals from equipment and filter media companies, operating companies, universities, engineering companies, as well as process innovators, startup groups, and consultants. Of course, I must also mention the used equipment dealers who provide a unique service to the industry. We see each other all over the world at conventions, conferences, customer sites, airports, and hotels. We share stories of the good and the "not-so-good" installations and always know we can do better.

My initial foray into this world of process filtration was at Pall Corporation under the tutelage of Dr. Pall and Mr. Abe Krasnoff. I learned the basics of filtration and how to really uncover process application details by markets and industries, solve problems, and then explain all of this under the guise of technical marketing. I later moved into another world of European process equipment with all of its mechanical intricacies and other types of process solutions with nutsche filter dryers, pressure and vacuum filtration systems, centrifuges, dryers, and all of the associated ancillary equipment such as solids handling, pumps, tanks, and reactors. Somewhere in the middle, I learned about sand and media filtration for cooling tower water, wastewater, and similar applications.

The question then for this acknowledgment page is how to thank everyone who has provided guidance, influence, help, and assistance. Throughout this guide, I have referenced various works as well as some others in the bibliography that I update and expand upon based on years of experience as well as customer contacts. Efforts were made to cite and credit within the guide anything believed to be material exclusive to that publication. Other articles and presentations, though not listed, from technical conferences were helpful sources of reference. Works, conversations, internal documents from colleagues, suppliers, friends, in some cases competitors who became friends, companies I worked for and with, magazine editors, and other experts in the filtration industry were also tremendously helpful in the development of this guide. They all share in its publication and to everyone please accept my sincere and heartfelt thanks. I'm sure as you read this guide you will recognize your contributions.

I am also grateful to Jon Worstell who provided the "push" to get this book under way by recognizing me as the right person to make this foray into penning a book collecting together current filtration expertise.

I would also like to thank BHS-Sonthofen GmbH, the parent company of BHS-Sonthofen Inc., the CEO and Chairman Dr. Christof Kemmann, and Dennis Kemmann, the Managing Director, for their trust in my judgment over the past 14 years as we have developed and grown the BHS business worldwide.

Lastly, my wife, Michelle, and my children, Jason and Bryan, as well as my extended family and friends have, of course, heard my stories and have been involved for over 30 years in the field of filtration just by being a part of my life. Sometimes they understood what I did and other times not, but they were always supportive as I tried to solve a process problem or come up with a new way to explain to the market a process solution. To all of them, thanks and hopefully you had as much fun as I did.

CHAPTER 1

Introduction

Solid-liquid filtration may not have the glamour of investigating blackmail and burglary, extortion and espionage, or murder and mayhem. Yet, process engineers can still learn many things from Sherlock Holmes and Dr. John Watson. While Sir Arthur Conan Doyle's fictional characters may never solve real-life process filtration problems, they prove time and again that there is no benefit to jumping to conclusions. The duo's sleuthing also benefits from working together to recreate events. Often Holmes talks through his theories to Watson—only then do gaps and inconsistencies become apparent. Additionally, Holmes and Watson apply the problem-solving skills such as the occasional silence, employing distancing and learning to tell the crucial from the incidental (Konnikova, 2013).

This practical guide to solid-liquid filtration is intended to better enable the inner Sherlock (or Watson) of engineers. Most university curriculums do not cover solid-liquid filtration, leaving many engineers clueless as to where or how to begin. This guide then will provide a framework to analyze and think about process filtration problems. It is not a "how to book" explaining in detail how to conduct tests or how to scale up from data; that is the purview of the equipment suppliers with deep knowledge of their individual technologies.

The information in this guide, gathered from over 30 years of experience, instead covers basic principles and mechanisms of filtration, filtration testing, including filter aids and filter media, types of filtration systems, selection of filtration systems, and typical operating and troubleshooting approaches. General applications and tips for process filtration are also included to better enable "idea-generation" for process engineers when analyzing filtration for an operating bottleneck issue or a new process development problem.

This chapter serves as an introduction to the fundamental concepts and terms process engineers need to understand. An overview of filtration, filtration equipment categories, principles and mechanisms, filter media, coagulation and flocculation, surface charges, and filter rating systems is initially explored in this chapter.

FILTRATION OVERVIEW

Process engineers really do have to employ their own kind of sleuthing, for selecting the right type of liquid/solid separating equipment is not a simple task. The wide range of equipment that might be employed and the at times illogical solutions required can cause further confusion among those who have no intimate knowledge of the processes. A further complication is that there are only two basic principles of liquid/solid separation, which means that equipment that might not be the best choice can still be made to work although, of course, sacrificing efficiency. This requires one to be extra careful when working with suppliers who only offer one type of equipment. The solution they are offering may, perhaps, be "made" to work, but may not necessarily be the smartest choice. For these reasons it is useful for engineers confronted with a separation problem to carry out their own basic evaluations of possible or probable solutions and to, above all, establish which routes not to follow.

It is hoped that this guide will enable the engineer to arrive at one, two, or three types of equipment that have an above-average chance of being right for the job, while also helping to eliminate those which are unsuitable. Having narrowed the field of options, the engineer has to calculate the strengths of weaknesses of each system through practical field pilot tests of actual machines and speaking with and visiting, if possible, the provisionally selected equipment's users.

FILTRATION EQUIPMENT CATEGORIES

There are several approaches to categorizing solid-liquid filtration equipment. As a result, when beginning a project, the opening question is how best to begin.

First, the engineer must decide whether the process is a batch or continuous process. This is not as simple as it sounds. For example, while the reaction/precipitation can be batch or continuous, the solid-liquid equipment can also be batch or continuous. Further, it is necessary to examine the downstream process and determine whether this process is batch or continuous. Most often, a continuous process is more efficient. However, other ideas must be considered such as time in the reactor, crystal sizing/breakage, solids handling (or alternatively, whether the process can handle a concentrated slurry rather than solids), drying time, and other parameters. In summary, a review of the entire process is necessary before determining how best to begin.

Having reviewed the process, the next factor to consider is what type of discharge is required: batch or continuous, dry solids, wet solids, or concentrated slurry. Further, the engineer must determine if the solids are the product or the liquid is the product or possibly both components are required. Besides these questions, another decision impacting any cost analysis is whether this is a manual or automatic operation. For instance, are the solids directly discharged to the downstream process, moved by conveyors, totes, or intermediate bulk containers (IBCs), or are other systems employed? Even with this information gathered, there is still more initial information needed.

Finally, the engineer must examine the amount of solids to be filtered. For example, does the process slurry contain "high solids," which could be up to 50–60% solids, or low solids, such as 2–5% down to trace amounts in the parts per million (PPM) level.

With all the above information in hand, the engineer can finally determine the best starting point. Just as Holmes and Watson agree that checklists and guides are important for problem solving, there are also several decision guides that have been published over the years. Davies (1965) proposed a scheme for selecting solid-liquid filtration equipment based upon cake dryness required, cake washing, filtrate clarity, and crystal breakage (Figure 1.1) (cited in Carpenter, 2013). Chase and Mayer (2003) offered a different approach looking more closely at the process including pretreatment, solids concentration, solids separation, and posttreatment (Figure 1.2). While these types of guides may be cumbersome, they do provide a good overview of what types of equipment are possible.

PRINCIPLES AND MECHANISMS

The two basic principles of solid-liquid filtration are to either separate liquids from solids or filter solids from liquids. This would suggest employing one of two basic solutions:

1. Either the solids have a tendency to go one way and the liquid the other way (i.e., separation), or

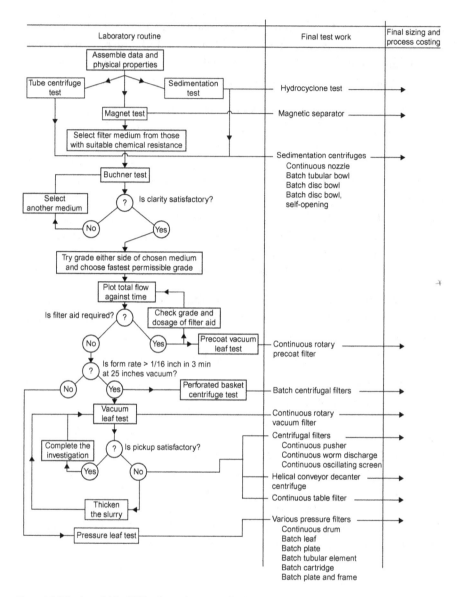

Figure 1.1 Selecting solid-liquid filtration equipment per Davies.

2. One must find a hole smaller than the solids which one wishes to capture (i.e., filtration).

Yet in spite of there being only two principles, there are over 100 different types of equipment available, many with their own variations.

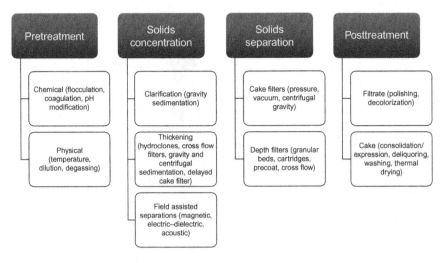

Figure 1.2 Chase and Mayer's approach to separation.

As this guide focuses on solid-liquid filtration, suspended solids are removed from liquids either on the surface (cake filtration) or within the depth of the filter medium. The depth of the filter media can be the filter media itself, the cake, or the filter aid. Regardless of the surface or depth filtration there are three mechanisms for removal: inertial impaction, diffusional interception, or direct interception.

Inertial Impaction

Particles in a fluid have a mass and velocity and hence have an associated momentum. As the liquid and entrained particles pass through a filter media, the liquid will take the path of least resistance and will be diverted around the fiber. The particles, because of their momentum, tend to travel in a straight line and, as a result, those particles located at or near the center of the flow line will strike or impact the fiber and are removed (Figure 1.3). Generally, larger particles will more readily deviate from the flow lines than will smaller ones. In practice, because the differential densities of the particles and liquids are very small, this mechanism is less effective in liquid filtration.

Diffusional Interception

For particles that are very small (i.e., very little mass), separation can result from diffusional interception. In this mechanism, particles are in collisions with the liquid molecules. These frequent collisions cause the particles to move in a random fashion around the fluid flow lines.

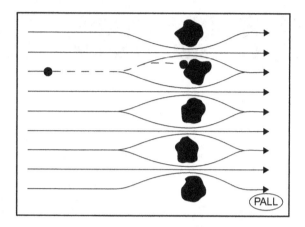

Figure 1.3 Inertial impaction.

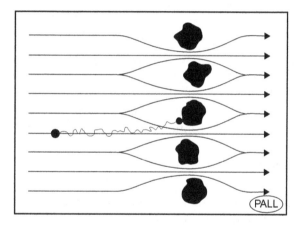

Figure 1.4 Diffusional interception.

Such movements, observed only on a microscopic level, are called Brownian motion. Motion causes the smaller particles to deviate from the fluid flow lines thus increasing the likelihood of the particles striking the fiber surface and being removed (Figure 1.4). However, as with inertial impaction, this mechanism has only a minor role in liquid filtration.

Direct Interception

Direct interception is equally effective in liquids and gases and is the predominant mechanism for removing particles from liquids. In the depth of the filter medium (medium itself, the cake, or the filter aid)

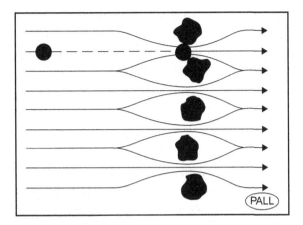

Figure 1.5 Direct interception.

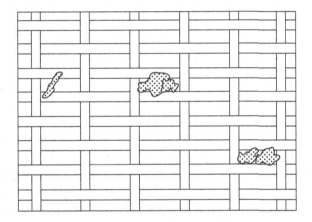

Figure 1.6 Particles striking a pore in direct interception.

one observes not only a single fiber or structure but also a rather tortuous path. The tortuous path defines the pores or openings that will remove the solids (Figure 1.5). Particles smaller than the pores can also be removed when they bridge across the structure or when two or more particles strike a pore simultaneously (Figure 1.6).

While these mechanisms are operable, the relative importance and role of each changes. Both inertial impaction and diffusional interception are much less effective with liquids than with gases. Since the density of the particles will be typically closer to that of the liquid than to that of the gas, deviation of a suspended particle from the liquid flow line is much less and thus impaction on structure of the medium is less likely. Moreover, impaction to the surface of the filter media is not followed by adhesion of the particles to the surface of the filter media. Diffusional interception in liquids occurs only to a very limited extent because Brownian motion is not nearly as pronounced in liquids as it is in gases.

FILTER AIDS

In some cases, to help the filtration process, filter aids are used to change the solid phase of the material. Filter aids range from proprietary products like diatomaceous earth or expanded silica to randomly selected crystalline materials. In general, they are only useful in filtration processes, although in specific cases they can assist settling if the suspension solids have a tendency to adhere to or impinge on the filter aid crystals. Diatomite, perlite, and cellulose are the most widely used porous media (filter aids) in dynamic process filtrations, with a high percentage of fine filtration applications using diatomite (Sulpizio, 2013).

Diatomite
Diatomite is obtained from diatomaceous earth. This sediment is greatly enriched in silica in the form of the siliceous diatoms (a diverse array of microscopic, single-cell algae of the class Bacillariophyceae). These diatoms are sufficiently durable to retain much of their structure through long periods of geologic time and through thermal processing. Diatomite products are characterized by an inherently intricate and highly porous structure composed primarily of silica, along with impurities of alumina, iron oxide, and alkaline earth oxides.

Perlite
Perlite is a naturally occurring volcanic glass that thermally expands upon processing. Perlite is chemically a sodium potassium aluminum silicate. After milling, a porous, complicated structure is present, but because its structure is not as intricate (or tortuous) as that of diatomite, perlite is better suited to the separation of coarse microparticulates from liquids having high solids loading. Perlite is lower in density than diatomite, which enables using less filter media (by weight). Perlite and diatomite are useful functional filtration components of depth filter sheets and pads.

Cellulose and Other Organic Media

Cellulose filter media is produced by the sulfite or sulfate processing of hard woods. Cellulose is characterized by its high aspect ratio, which enables it to precoat a filter media very easily. It is most often used in that capacity in combination with diatomite. Like perlite, cellulose possesses a less intricate structure than diatomite.

Other organic media includes potato starch particles, cotton linter, and polymeric fibers and flakes. These materials can help disperse diatomite in some systems or are specific to certain applications. An unusual mineral filter aid of organic origin is the ash from the combustion of rice hulls. This material has high silica content and a residual carbon char and has been found to be useful in waste treatment and stabilization of hazardous materials.

Filter aid can be used as a precoat or a body feed. As a precoat, the filter aid protects the filter media against the penetration of unwanted solids and premature blinding of the media. In practice a combination of the two approaches is most common. In all cases one must remember that this filter aid becomes part of the solids and that normally there is no practical way to separate them so they add to the amount of solids that must be disposed.

Selecting the appropriate filter aid depends upon the engineer's understanding of the specific application and filtration objective. For example, particle size, settling rate, solids density, and most important the solid's characteristics (granular, slimy, coarse, fine, etc.) are critical parameters to consider. How the solids to be removed will interact with the filter aid is another consideration. If the solids form an impermeable layer, then the thickness of the precoat does not matter since filtration will stop once the layer of solids is formed. If the solids are more fibrous and amorphous and can get into the depth of the precoat, then a thicker precoat will be beneficial.

In summary, there will always be a tradeoff of precoat, body feed, and the optimization of their usage compared to the blinding of the filter media as well as the filtration flux rate.

FILTER MEDIA

For the purposes of this guide, there are two main types of filter: media synthetic cloth or metal. The choice of media depends on filtration removal efficiency, process requirements, filter technology, characteristics of the solids and liquids, and other parameters (such as chemical and thermal resistance).

Media Synthetic Cloth

In terms of synthetic cloths, the materials can be polyester, nylon, polypropylene, PVDF (Kynar), PEEK, or fluoropolymers (such as ETFE, PTFE, E-CTFE, carbonized, and polyester). The media can then be segmented by the degree of openness: plain, square weaves have visible or nearly visible weave openings (normally larger than 200 microns) and closed weave filter cloth (from 1 to up to 200 microns). For the weave itself, media suppliers speak of weft and warp. The warp is the lengthwise or longitudinal thread in a roll, while weft is the transverse thread.

Media suppliers also talk about monofilament and multifilament. Monofilaments are the most concise and regular fabrics. The single strand threads are capable of exact detail. They wear well and solids are less likely to adhere. They are easier to clean and less likely to blind. The surface of the threads is less coarse, or more smooth and polished. Multifilament are fabrics that are fine strands twisted together into threads. The threads are coarse and have mild elasticity. Multifilaments are harder to clean and can sometimes trap solids. There are also filter media that can be both mono- and multifilament.

Lastly, there are different specifications for open area, thread size, and pore sizes. Percentages are given that indicate the relationship between the total open area of the mesh and the area that is covered by the threads themselves. A greater percentage of open area permits higher filtration rates. Greater thread size has more strength but diminishes the percentage of open area in the total mesh. The actual number of threads per square inch of fabric is always the same. Then, the media suppliers make it even more confusing with different descriptions and nomenclature.

Plain, Square Weave

These are the most basic open weave made with monofilament fibers. They have a simple over and under weave pattern with a straight flow path, high open area, and high permeability. They are easily cleaned but have a low stability in dynamic processes (Figure 1.7).

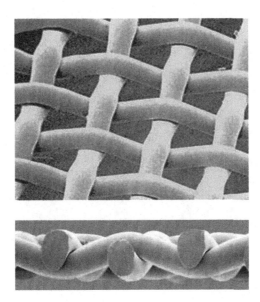

Figure 1.7 Plain, square weave.

Twill Weave

These are also monofilament and, for example, can be defined as 3/1 twill. In this example, warp threads go over 3, then under 1 weft thread for a nonsymmetric weave. This weave has very dense patterns and good strength and durability and is considered a filtration workhorse. They are often calendered (another new nomenclature) to adjust pore size, air permeability, and improve the solid's release. Calendared monofilament is a nonwoven textile that has been rolled between cylinders to control thickness and smoothness (Figure 1.8).

Plain, Reverse Dutch (PRD)

These are also monofilament fibers with higher warp thread count than weft. The warp yarn diameter generally is 2/3 weft yarn diameter. The media has a very tortuous flow with symmetrical weave and a very high flow rate for a given pore size (Figure 1.9).

Double Layer Weave (DLW)

These can be monofilament or monofilament/multifilament fiber combinations. The media is durable and robust with high strength in both directions. They have a smooth filter cake side and backside flow feature (Figure 1.10).

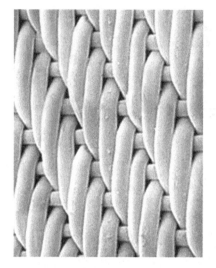

Figure 1.8 Twill weave.

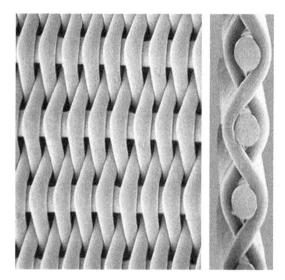

Figure 1.9 Plain, reverse Dutch.

Metal Media

The second alternative is metal media. Metal media can be single layer or multilayer and can be different types of stainless steel and alloys, such as Hastelloy, Inconel, nickel, Monel, titanium, and others. For single layer metal mesh, the descriptions are similar to the cloth weaves.

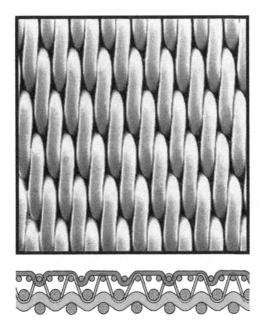

Figure 1.10 Double layer weave.

Plain Weave

Plain woven is the most common wire cloth weave. A weft wire passes alternately over and under each warp wire, and each warp wire passes alternately over and under each weft wire (Figure 1.11). Both warp and weft wire diameters are generally the same.

Twilled Weave

Each weft wire alternately passes over two, then under two successive warp wires, and each warp wire passes alternately over two and under two successive weft wires, in a staggered arrangement (Figure 1.12). Twill weave is normally used to allow a heavier-than-normal wire diameter in association with a given mesh.

Plain Dutch Weave

Plain Dutch wire cloth weave has similar interlacing to plain weave, except the warp wires are larger in diameter than the weft wires. While the warp wires remain straight, the weft wires are plain woven to lie as close as possible against each other in a linen weave forming a dense strong material with small, irregular and twisting passageways that appear triangular when diagonally viewing the weave (Figure 1.13).

Figure 1.11 Plain weave in metal media.

Figure 1.12 Twilled weave in metal media.

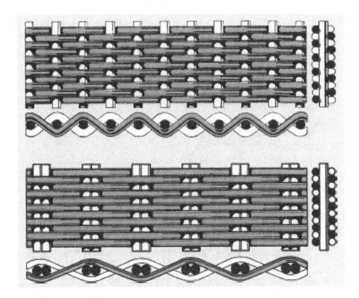

Figure 1.13 Plain Dutch weave in metal media.

The Twilled Dutch

This weave is similar to Plain Dutch, except that the weave is twilled, allowing a double layer of weft wires. There are no "straight-shot" apertures through the mesh, and the filtrate follows a tortuous path to pass through the wire cloth (Figure 1.14).

In terms of multilayer, these can be sintered or un-sintered. With un-sintered material the interlocking of the weave is the only strength factor that prevents wire movement. Sintering makes permanent the geometry of the original weave, and therefore the pore size is fixed.

Sintered metal media is constructed of multiple layers of wire mesh and are designed for precise controlled porosity, uniform pore sizes, and distributions. The laminates are permanently bonded under precise diffusion bonding (sintering) conditions. The standard designs can be two, three, or five layers. The middle layer is normally the removal efficiency while the top layer is a protective layer and the bottom layers are for draining.

COAGULANTS AND FLOCCULANTS

As discussed earlier, filter aids can be used to alter the material phase of the solids. In some situations, to aid in the filtration

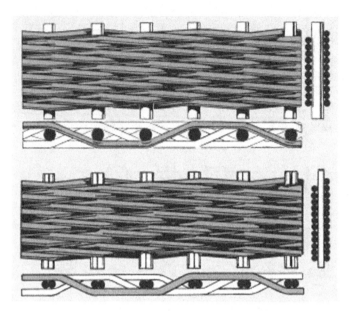

Figure 1.14 Twilled Dutch weave in metal media.

process, it is necessary to change the effective particle size of the solids. In these cases, coagulation and flocculation processes are used. These can only be used if the process itself can tolerate added chemicals. The suspended particles in a slurry will vary in size, shape, and density. Coagulation and flocculation occur in successive steps intended to overcome the forces stabilizing the suspended particles, allowing particle collision and growth of solids to larger particles to help filtration. If one step is incomplete, the following step will be unsuccessful.

The chemical supplier in consultation with the filter supplier will likely determine the types of coagulants used. Chemically, coagulant chemicals are either metallic salts (such as alum) or polymers. Polymers are man-made organic compounds made up of a long chain of smaller molecules. Polymers can be either cationic (positively charged), anionic (negatively charged), or nonionic (neutrally charged).

Coagulants are usually fed into the water using a gravimetric feeder or a metering pump. A gravimetric feeder feeds dry chemicals into the water by weight. A metering pump feeds a wet solution (a liquid) into the water by pumping a volume of solution with each stroke or rotation.

The first step destabilizes the particle's charges. Coagulants with charges opposite those of the suspended solids are added to the slurry to neutralize the negative charges. Once the charge is neutralized, the small, suspended particles are capable of sticking together. The slightly larger particles formed through this process, called microflocs, are invisible to the naked eye. The water surrounding the newly formed microflocs should be clear. If it is not, all the particles' charges have not been neutralized, and coagulation has not been carried to completion. More coagulant may be needed.

Following the coagulation, a second process called flocculation occurs. Flocculation, a gentle mixing stage, increases the particle size from submicroscopic microfloc to visible suspended particles. The microflocs are brought into contact with each other through the process of slow mixing. Collisions of the microfloc particles cause them to bond to produce larger, visible flocs. The floc size continues to build through additional collisions and interaction with the polymers added. Once the floc has reached it optimum size and strength, the slurry is pumped to the filtration system. It is important to select the correct pump for the flocked slurry to insure the pump's shearing forces do not destroy the flocs.

This section provides only a brief introduction to these processes. Two examples of the use of coagulants and flocculation will be described in greater detail in Chapter 5.

SURFACE CHARGES

Surface charges on particles can affect the rate and extent of the filtration process. The magnitude of the actual charge is not easily measured, but the effective charge of the solids in the slurry can be characterized by the solid's zeta potential. What does this mean for filtration? At times, maximum repulsive forces are beneficial to keep the particles discrete and prevent them from growing into larger particles and settling. In other cases, the opposite effect is desirable as by removing the forces the particles are allowed to grow and settle. The real answer depends upon the process conditions and application.

It is important to note that the zeta potential can vary due to the pH and ionic species strength in the slurry. For example, the zeta potential can be lowered by adding non-adsorbed electrolyte ions to the bulk slurry to promote agglomeration of the solids. Of course, these parameters can only be modified if they are compatible with the process.

The zeta potential of solids is generally more important for the filter media suppliers of filter cartridges. Generally, most solids are negatively charged. Yet, the filter media can be positively charged to enhance the particle capture and adherence. The intensity of the charge of both the particle and the fiber is critical. Normally, as the charge increases and the particle size decreases, the capture efficiency will increase. The obvious effect is the ability to remove very fine particles by a media with large pore sizes and very low-pressure drop and high dirt capacity. Throughout, however, it is necessary to remember that the charge or zeta potential is often unstable and can influenced by pH, humidity, ionic strength, and more. Therefore, the predictability of the removal efficiency is poor and should always first be verified by testing.

FILTER RATING SYSTEMS

While various rating systems have evolved over time to describe the filtration capabilities of filter media, there is no generally accepted rating system. Unfortunately, this tends to confuse the filter user resulting in misapplication of the media. This section will describe the four commonly used rating methods (nominal, absolute, beta ratio, and air permeability). But, to reiterate this guide's general tenet: testing of the application is the first step for selection.

The rating of the media is always reporting "clean" removal efficiency. Remember that the actual removal efficiency will depend on the filtration mechanism of surface or depth filtration as well as factors including particle size or shape. If a cake is building up, then the cake becomes the filter media and removes finer particles. If filter aids are used, then these are the guiding parameters. Therefore, as with all filtration processes, there is no one, clear answer of what media to specify. Nevertheless, an understanding of the rating methods can help with the problem-solving process.

Nominal Rating

The nominal filter rating is an arbitrary micron value given to the filter by the manufacturer based upon removal of some percentage of all particles of a given size or larger. It is rarely well defined and not reproducible. Ultimately, these ratings have little to no value. In practice, a contaminant is introduced upstream of the media and the effluent (filtrate) is analyzed. Generally, the definition is that 98% by weight of a certain size has been removed and 2% has passed downstream. Keep in mind, though, this is weight removal and not particle size (or count). By modifying the conditions such as using high upstream concentrations or different removal efficiencies, the ratings can change. It is not unusual that particles as large as 200 microns will pass through a nominally rated 10-micron filter.

Absolute Rating
Another common rating for filter media is the absolute rating. An absolute rating gives the size of the largest hard spherical particle that will pass through the filter or screen under specified test conditions. This rating can only be assigned if the media (pore openings) is fixed in place such as by sintering or fixed-pore bonded fibers. Essentially, this is the size of the largest opening in the filter. There are several methods used by media manufacturers including glass bead challenge testing, single pass and multipass testing, and bubble point tests. In summary, absolute ratings are better for representing the effectiveness of a filter over nominal ratings.

Beta Ratio
The Beta ratio is a simple rating system based upon total particle counts. This is the ratio between the per unit volume number of particles above a given size in the influent (upstream) of the media suspension to the same parameter in the effluent (downstream) of the filter media. Percent removal can be determined at the given particle size as:

$$\%\text{Removal} = \frac{\text{beta} - 1}{\text{beta}} \times 100$$

It follows that the higher the value of the beta ratio, the more particles of the specified size, or greater, are removed by the filter media. Usually, a Beta ratio = 10,000 can be used as the operational definition of an absolute rating. Typical ratios are shown in Table 1.1.

Air Permeability
Air permeability is defined as the flow rate of air per unit area at a given differential pressure and is normally expressed as cfm/ft^2 at 0.5 in. water gauge. Typical ratings can be from 2 to 2000. The result

Table 1.1 Typical Beta Ratio Ranges	
Beta Ratio	% Removal Efficiency
1	0
2	50
10	90
100	99
1000	99.9
10,000	99.99
100,000	99.999

will be the relative pore size, so a small number would provide finer removal efficiency. ASTM D737-04 is the standard test method of air permeability of textile fabrics.

Construction factors and finishing techniques have an effect upon air permeability by causing a change in the airflow paths through a fabric. Hot calendaring can be used to flatten fabric components, thus reducing air permeability. Fabrics with different surface textures on either side can have a different air permeability depending upon the direction of the air's flow.

For woven fabric, yarn twist also is important. As twist increases, the circularity and density of the yarn increases, thus reducing the yarn diameter and increasing the air permeability. Yarn crimp and weave influence the shape and area of the interstices between yarns and may permit yarns to extend easily. Such yarn extension would open up the fabric, increase the free area, and increase the air permeability. Finally, increasing yarn twist also may allow the more circular, high-density yarns to be packed closer together in a tightly woven structure with reduced air permeability.

Having offered this concise introduction to filtration, filtration equipment categories, principals and mechanisms, filter media, coagulation and flocculation, surface charges, and filter rating systems, this guide next considers filtration testing.

CHAPTER 2

Filtration Testing

Testing plays an important role in filtration and is the key for selecting the most suitable filtration technology for the individual solid-liquid separation task. Although there are only limited theoretical backgrounds available, and even specialized engineering education at universities leaves many theoretical questions open, it is beneficial to have at least a minimal understanding of the theory of filtration itself. Identifying the role of each influencing part, the process engineer receives a potential tool to work with when it comes to understanding testing's findings and developing a path forward.

This guide does not detail every step in preparing and conducting filtration tests. Every test should be individually designed on a case-by-case basis. Furthermore, it is generally assumed that the testing and engineering skills to follow the test procedure are mandatory, and standard operation procedures (SOPs) for the test equipment are available from the equipment's manufacturer. However, as Sherlock Holmes often warns Watson not to jump to conclusions, this is also one of the biggest risks process engineers face during the testing process.

From experience and for the benefit of engineers, some overview observations are necessary:

- Don't stop testing just because the first results suit your target.
- Don't accept samples without verifying the parameters in the description.
- Never change more than one parameter at a time.
- One result is no result. Verification is a must.
- Take a break and check the conformity of the results before you call it a day.

That said, this chapter addresses filtration theory background and practice and slurry characteristics before discussing particle shape, size, concentration, and measurement as well as pressure and vacuum testing.

FILTRATION THEORY: BACKGROUND AND HOW TO USE IT IN PRACTICE

With a little basic filtration theory, the process engineer can make great strides in selecting the best processes for each unique situation. Although the process engineer might have good reasons to select a continuous or batch process, in the initial view due to the overall process (including the upstream and downstream technologies) the product's slurry characteristics very often do not match the engineer's first choice. Filtration testing is a method to identify slurry characteristics. As these characteristics can only be influenced within limits, but have a great impact on overall process, identifying them as early as possible in the planning phase is beneficial.

The filtration technology differs among three major cases for solid-liquid filtration only: (i) cake building, (ii) non-cake-building filtration as dynamic filtration, and (iii) deep-bed filtration (Schubert & Rippberger, 2003).

Cake-building filtration is defined as removing solids from a suspension by the use of a filter media through which the filtrate is passing. The filter media is starting the cake-building process. Once initiated, the cake itself functions as a solids trap for the remaining filtration process. No separation in the filter media takes place; the solids are either collected in the cake or they pass into the filtrate. The built cake will later be removed from the filter cloth (and therefore needs to have a minimum height of at least several millimeters).

Suspensions that do not allow building a practicable cake height within an economical time typically have only a low solid content and/or are very fine dispersed. For these slurries a dynamic filtration (i.e., cross-flow filtration) can be a solution. In this filtration principle, the solids are also separated at the surface of a filter media, but as soon as a cake is built, it is taken away by a high-speed cross flow (of slurry), maybe supported by a back pulse. With this method, it is possible to generate a clear (solid-free) filtrate, but only a thickened sludge of solids, not a cake.

As a third alternative, deep-bed filtration is used. This method captures the solids of the suspension within the filter media, such that no cake is built on top of the media. The principle of deep-bed cake filtration occurs with large particles that form a very permeable cake such that as the cake builds up to over several inches, the flow of mother liquor remains high and/or constant. Another possibility occurs using a filter aid such that the filter aid forms a precoat and the particles are amorphous such that they are removed within the depth of the precoat layer.

Many parameters have an influence on the filtration process. These include, but are not limited to:

- form and size of particles
- particle size distribution (PSD)
- agglomerate building behavior
- deformability
- compressibility
- viscosity
- solid content
- zeta potential
- pressure

While all of the above may not be known for all filtration applications, the final target is always to find a theoretical approach together with a practical method of testing.

The filtration theory upon which cake building and deep-bed filtration is based is Darcy's law. The law describes the flow of fluids through porous materials (Darcy, 1836). Dynamic or cross-flow filtration is not part of this guide and therefore is not discussed. In the VDI-Guideline 2762 for filtration, the theoretical development is more detailed; however, the main focus for this guide is the testing results and how the engineer can use testing data to make correct process decisions.

Formulated by Henry Darcy, based on the results of experiments on water's flow through sand beds, Darcy's law forms the scientific basis of fluid permeability. A practical equation was developed from Darcy's law based upon the following assumptions:

- The built cake is not compressible.
- The pressure during the cake building is constant.
- The filtrate is clear and all solids from the suspension do end up in the cake.
- The resistance created by the filter media is negligible compared with the cake resistance.

Experiences have shown, in many cases, considering the previous assumptions, that a single equation can be used (Nicolau, 2003) (Figure 2.1).

Most cases of everyday filtration testing can be described with Darcy's equation. The first focus of the relationship is on the definitions of the various parameters:

1. All parameters on the right side are only a square rooted relation to the sizes on the left side (under proportional relationship).
2. The pressure and time are increasing values (more pressure leads to more capacity).
3. All others (viscosity, solid content, etc.) are decreasing values (more solid content leads to less capacity).
4. Alpha is a sum of all "unknowns" such as PSD, porosity, solid's shape and size, etc.

As the filtration tests for a specific slurry are usually made with the same slurry sample, a secondary assumption is that the viscosity and solid concentration can be taken as consistent for this testing. This results in another step of simplification and finally ends in (assuming that all test parameters are kept constant mainly pressure, temperature, and sample preparation) the equations demonstrated in Figures 2.2 and 2.3. The two most simplified equations (Figure 2.2) describe the relationship of the specific filtration flow (V/A) to the filtration time ($\sqrt{t_F}$) as a square rooted relationship.

$$\frac{V}{A} = \sqrt{\frac{2 \cdot \Delta p}{\alpha \cdot c_m \cdot \eta_f}} \cdot \sqrt{t}$$

Figure 2.1 Darcy equation.

$$\frac{V}{A} = x \cdot \sqrt{t_F}$$

Figure 2.2 Simplified Darcy equation showing the relationship between filtration flow and time.

$$\frac{V}{A} = y \cdot H$$

Figure 2.3 Simplified Darcy equation showing the relationship between filtration flow and cake height.

Figure 2.3 describes the relationship of the build cake height (H) in relation to the filtration flow (V/A). To allow for practical use, the entire unknowns are summed up in the two factors X and Y. For everyday test work these two factors can be determined for each individual case by several bench top tests. With this in mind, the process engineer now can make theoretical assumptions by altering the parameters without needing all details at hand.

Let's take an example to get a better idea of what is behind the theoretical approach:

The first test has given a filtration time of 60 s, 20 mm of cake height and a specific volume of 100 l/m^2. The production filter has to be sized for 15 m^3/h.

The first scale up results in a filter area of 2.5 m^2 (15,000 l/h/ 6,000 l/m^{2}h). Filter cake height and specific throughput as well as filter speed are within the standard range of a potential filter, so technically there is no need to alter anything; however, it is worth having an economical look at the results.

The process engineer can reconsider his or her test data assuming a usable filter type of this size costs 100%. Following the given equations, a potential gain of filtration speed can be taken from the reduction of the cake height. This has a linear relation to the specific flow, but the filtration time has a square rooted relation. This means every modification of the cake height has an overproportional effect on filtration time.

In this example, the cake height can be reduced to 50% of its first tested value (from 20 to 10 mm, assuming this is still OK for the production filter in our example). The filtration time is then reduced to 25% of the initial time $(1/2)^2 = (1/4)$. Thus, the filtration time (for 10 mm) is only 15 s instead of 60 s. The filtered slurry is also reduced, but only to 50% ($= 50$ l/m^2). So, all in all, the reduction in cake height is a gain in throughput 50 l/m^2 $\times$ 240 s/h $=$ 12,000 l/m^2/h. Doubling the filtration flux rate halved the required filter area, which should have a positive effect on costs as well. The same effect is also evidenced with increasing cake height. For example, increasing the cake height by 50% requires 225% of filtration time in total and reduces overall performance. As a consequence, very low cake heights (fraction of a mm) achieve the highest (theoretically) flux rates. But there are practical limits. Not only does the production filter require a minimum cake height to work, but also the filter cloth resistance is then gaining more importance the smaller the cake gets; therefore, the whole scenario only works if the filter media resistance is negligible compared to the cake resistance itself (Tichy, 2007).

In summary, during the testing and observing of the cake thickness, there are additional impacts such as cake compression, etc. that may become more influential in the process. For large theoretical scale up from empirical data, careful analysis is necessary. In any case, it is recommended to test under operational conditions as close as possible to the production conditions. If possible, the next step in our example would have been to test a cake height of 10 mm to see how good the theory matches the practical test data. Remember, don't jump to any conclusions.

SLURRY CHARACTERISTICS

In testing filtration theory, it is important to use a representative sample. This refers to a sample as close as possible to the real production product. Yet, the specific characteristics of a slurry under the point of filtration view are not always obvious. The list of parameters already discussed in the chapter was quite extensive. In testing, in many cases, only a few of the parameters are measurable, but the more data the process engineer has available the better. Although for the first tests, the pH value, temperature, particle shape, or size distribution are not really needed right from the beginning, they are quickly measured parameters and can complete the picture of the suspension. It is obvious that solid content and viscosity impact the filterability. Remembering the filtration equation, these parameters are very important. Influence from variation of viscosity, for instance, has significant effects.

In interpretating the equations in Figures 2.1–2.3, the process engineer working with filtration tests must remember three key points:

1. The higher the solid concentration C_m is, the less slurry can be filtered in the same time.
2. The higher the solid concentration, the higher the cake height (with the same amount of slurry).
3. For the same cake height, the filtration time has a reverse proportional relationship to the solid concentration. This means diluted slurry takes more time while pre-thickened slurry takes less time to achieve the same cake height.

How do these guiding principles help a process engineer with a diluted suspension that takes a long time to filter and ends in uneconomical large filter size? With these tenets in mind the engineer knows the answer may be the use of a pre-thickener (static thickener, hydro cyclone, or settling tank among others) to raise the solid concentration to a level such that an economical cake-building filtration becomes possible.

Viscosity has the same reverse impact on filtration speed. However, it has no additional effect on the cake building, similar to increases in the solid concentration. However, a reduction of the viscosity always benefits the whole filtration process. Not only will the flux rate increase, but also, as a side effect, the surface tension of the liquid in the capillary is reduced, directly helping to mechanically reduce the cake moisture at the end.

As for many liquids, the viscosity is mainly influenced by the temperature. Thus, it is important to check the potential for this as well. As every change in parameters can have further effects, temperature evaluations always prompt reverification with the upstream and downstream process and checking availability of energy in the process.

Modification of the chemical parameters, such as solubility of dissolved materials, can also be evaluated. In some cases even the solids to be filtered could dissolve as a result of adding a few degrees of temperature. Unfortunately, this typically does not happen suddenly nor for all solids at once. Yet dissolving just a few of the fines (they go into solution first) can have a huge impact to overall PSD, thus improving the filtration behavior.

So, practically speaking, the first choice is always to use the temperature given by the process (mainly upstream) for testing, making sure to keep the temperature constant during the entire testing. Especially at high temperature levels this can be a challenge, and in some cases even become an obstacle that jeopardizes the whole testing. Then, a theoretical correction of the viscosity can be a path forward (don't forget to mention every detail in the report and make transparent any alterations to the test procedures).

In making decisions and testing, the process engineer should also consider the other half of the slurry—the particles. These relevant parameters are examined in greater detail later.

PARTICLE CHARACTERISTICS

The characteristics of suspensions are not only caused by the liquid phase but also by the particles, the other half of a slurry. The solids can be of crystalline nature or amorphous, which means their structure is not really defined. They can also be organic (e.g., cell debris), fibrous, inorganic, compressible or incompressible, generate agglomerates or not, or may have a zeta potential or not ... There are many possibilities.

Crystals, at first glance, are beneficial for filtration, as they have a regular structure and are usually incompressible. Unfortunately, this is only true for the single crystal. Sometimes, a collected sum of single crystals in a filter will not necessarily end in an incompressible filter cake. According to Ives (1975), this is very much influenced by the shape of the crystals (or solids) and further parameters of the suspension.

Ideal solids are regularly shaped, potentially round as glass beads, and of one size only. Then, the built cake is always and everywhere identical, no matter how the individual solid is arranged while in cake structure. The commonly used filtration pressures are not high enough to change the individual structure of a glass bead, and the consistency of the cake prevents it from compressing when under pressure (otherwise known as incompressible and nondeformable).

Knowing round and hard is good for the filtration immediately suggests what solids are not as beneficial: those that are soft and flat or extremely tall (diameter length ratio >3). Needle-shaped solids are shear sensitive (their risk of breaking equals a resulting reduction of particle size) and build irregular filter cakes, often easily deformable felt-like structures. While these options are not necessarily bad for slurry filterability, because these cakes are very sensitive, it can become quite a headache to generate identical parameters and therefore comparable results when transferring test data to production machines.

As for other types of solids, those which are flaky, scale-like solids (some minerals, graphites) are more difficult because they are prone to building impenetrable cakes if treated with too high pressure. While organic cells are very often round, they are even less resistant against pressure. They are quite easy to deform and a tendency to collapse when over pressured. Many amorphous cakes do sooner or later react on pressure loadings as well.

These cases demonstrate the restricted use of filtration theory. Ultimately, a lot depends on the parameters of each individual case. Very often in practice the most ideal operational pressure is not very high (1 bar g or less), as then the negative effects of deformation, etc. do not become dominant in the cake resistance.

An easy way to verify the type of solids is a sample check under the microscope. If possible, the original suspension should be checked under the microscope. In observing the behavior of the solids the engineer might consider: Do the solids tend to build agglomerates or stay on their own? How is the distribution? The main information needed is the structure of the solid—is its shape needle, potato, snow crystal, or even fibrous?

Along with shape, it is crucial to understand particle size. For the PSD of the solids, suitable measurement devices should be used. Cake-building filtration typically operates within a 1−200 micron range. In fact, the majority is between 10 and 100 micron (average particle size). Under the one micron size, the solids are difficult to filter with cake-building technology. On the other hand, larger solids, those greater than several millimeters, can be handled by sieves, continuous centrifuges, or settling devices.

Particle size analysis can be accomplished using different techniques. The simplest and cheapest is sieving, in which dry solids are run over a variety of exactly defined sieves, which allow weighing back the mass of the solid's hold on a certain sieve size given as x% of 43 micron sieve or x grams between 43 and 60 microns. The sizes are staggered and end with the 43 microns on the low end and go up to 200−300 microns. For many suspensions, though, this technique is inadequate as the information on the lower end needs more detailed verification. These verifications are done with laser diffraction or optical counting methods. As the PSD gathers some of the most important information needed to characterize a suspension, the techniques employed must be suitable for the individual product. Many labs are available to run a particle analysis for little money per test. Unfortunately, results are not always easy and fast to get. Nevertheless, in many cases, this is most essential to understanding the whole suspension, especially when test results are going to be verified in production. Most filtration problems can be traced back to the root cause of differences in PSD.

To make things even more complicated, it is the *D* (distribution) that matters most. Sometimes, the average distribution of the particle size is already known, and this might not have changed at all, or could have become bigger (and bigger should be better). As a result the assumption may be that this average PSD can be used. Regrettably, this is incorrect. The average grain or particle size is only a hint that is not yet detailed enough.

The PSD is normally listed in a table with absolute mass per segment (<1, 1–2, 3–5, 6–10, 10–20, 20–30 microns, etc.), or % of mass per segment, or in a table representing both with dual scale. These are often accompanied by a graph giving a visual idea of the distribution. Taking a PSD from the lab test sample and comparing it with the production suspension can be very helpful to understand why filtration results are different. If the particle size is different, there are more fines or multiple peaks of shifted spreading or everything influences filterability. The graphs make it easier for engineers to gather impressions about the solid's behavior in a filter cake.

For instance, Figure 2.4 shows what would be considered an ideal distribution. The graph depicts a high peak somewhere in the middle and small tailings left and right, even shorter on the low end. Such a distribution would be nice to have. As the spread is not so big, the majority of the particles are all around the average size, so a cake built will easily become a quite homogeneous structure.

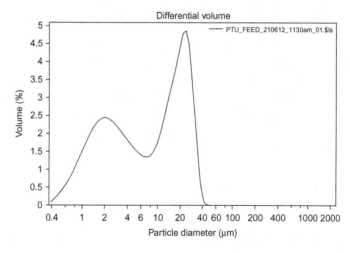

Figure 2.4 Ideal PSD with one main peak.

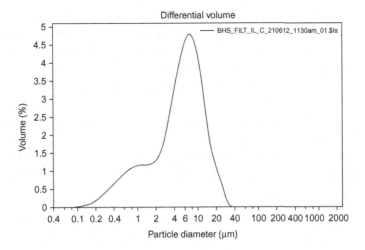

Figure 2.5 Multiple (twin) peak and wide PSD.

In contrast, Figure 2.5 shows a much different situation. Not only is the overall spread much wider, but it also shows multiple peaks. Typically, these graphs are using an exponential scale, so the gaps between the peaks are huge. Such a twin-peak distribution indicates an inhomogeneous cake, as the very fine particles (and there are quite a lot of these) tend to sit in the gaps between the rock-sized and much bigger ones. Or, even worse, the cake will become a segregated one, with the fines all going through to the cloth itself and the big particles building a separate cake on top of it. Now this behavior will suffer from an extensively increased resistance of the fines. The structure will be more similar to a PSD with a smaller peak (as to the left part until the first dimple) while the remaining larger particle distribution will have much less influence. It is obvious that although the average grain size might be identical, Figure 2.4's suspension will have a much better filtration rate than the suspension from Figure 2.5.

In dealing with suspensions, choices must also be made about pressure or vacuum testing. These methods are outlined later.

PRESSURE TESTING

Pressure testing is always the preferred method for products with elevated temperatures or super-saturated solutions. This method is also recommended for hazardous suspensions or any others that demand a completely closed or inert process. Besides the process benefit, it is normally better to use a much higher pressure difference for the filtration step in comparison to the restricted pressure difference in vacuum filtration technologies. The nature of this difference suggests suspensions that are expected to filter moderate or slow can be expected to be better processed with pressure filtration. Depending on the overall process, batch or continuous filter systems can also be foreseen. However, for continuous pressure filtration the available technologies are more restricted. Additional limits that should be taken into account prior to focusing on a certain filtration process are examined further in the following chapters of this guide.

Under the perspective of driving force, pressure filtration can easily offer three to five times higher rates than the vacuum filtration easily; the effect is the square root of the delta pressure. For example, comparing an average vacuum filter system's 0.5 bar g to a medium pressure filter with 3 bar g, the difference is 600%, which still would theoretically be good for a 2.4 times faster cake-building speed or filtration speed. Aiming for higher assuming it is better is not illogical, but the most practical filter types still do have pressure limits. Not only does the generation of higher pressure cost additional energy, but designing the filter for higher pressure could also add costs. Thus, the majority of pressure filter systems are in the 1−6 bar g range, and few high pressure systems (<25 bar g) are common. For everyday testing, up to 6 bar g is the pressure level of choice.

When, then, would the process engineer employ vacuum testing? This is answered later.

VACUUM TESTING

As reiterated throughout this guide, there are special considerations for every situation the engineer may confront in solid-liquid filtration. Vacuum testing is another area of differentiation. Fast filtering suspensions, fast settling solids, and nonhazardous products can each be processed in vacuum filter systems. As many filtration applications are neither explosive nor flammable and do not need special environment or operator protection, it is possible to use open filter systems. Potential candidates for open filtration systems include products from the mining industry, mineral processing, nonorganic products in basic chemistry, and the like. The filterability for the use of a vacuum system should be relatively faster than the pressure-suited products. Typically these products can build a cake within a few seconds, having applied a vacuum of approximately 0.5 bar. The normal practicable vacuum for most filters is in the range of 0.4−0.6 bar. Take into account that a further increase from 0.6 to 0.9 would only mean a pressure improvement of 50% resulting in a filtration speed effect of 22%. The achievement of 0.9 bar vacuum requires huge effort, so a vacuum around 0.5 bar is most common.

As filtration under vacuum only works with sucking air through the cake, this might have an impact on precipitation and fouling on the filtrate side. Especially saturated and hot suspensions are riskier than cold ones. It is essential to have an eye on the fouling effect, which can be verified by watching the filtrates carefully and checking the used filter cloths for signs of blinding during repeated usage. This caveat is not intended to reduce the use of vacuum filtration, but to remind engineers to keep this in mind during the testing.

Having expanded the engineer's understanding of theory and testing considerations, this guide will delve deeper into explanations of the types of filtration systems.

CHAPTER 3

Types of Filtration Systems

Filtration, pressure, or vacuum is the art of finding a filter media that allows the liquid to pass through while retaining the solids. The driving force may be gravity, vacuum, or pressure. A great deal of processing can often take place in the actual filtration stage, especially with regard to the conversion or exchange of the residual cake moisture, as the ultimate solid phase is always drier than that from gravitational systems.

The driving force, causing the liquid to exit while leaving the solids behind, can be divided into four groups:

1. Gravitational—Really little more than draining, this can be very useful to reduce large quantities to more manageable proportions. They can be batch operated or, more usually, continuous.
2. Vacuum—This is in many ways mechanically the simplest driving force available. Vacuum filters can be batch operated but are normally continuous. In general, the solid (cake) thickness can be controlled within close limits; there are few limits to materials of construction and some of vacuum filters offer the best solids washing possibilities.
3. Pressure—With or without compression this force obviously involves mechanical constraints. All pressure filters are batch-operated units with one exception—the rotary pressure filter. Cake washing can be excellent, and the final cakes are usually as dry as can be expected without heat input.
4. Centrifugal—This can sometimes offer a compromise between vacuum and pressure filtration. Centrifugal filters can be continuous in operation or can operate in an automated, continuous batch mode. The nature and behavior of the solids will play a great part in the success or failure of centrifugal filters. Cake washing can be good depending on the type and the behavior of the solids.

With these criteria in mind, the process engineer must make several basic choices. These are segmented by whether or not the filter will discharge the cake in a batch-wise or continuous manner. Within this categorization, there are a myriad of choices of filter technologies, designs, and manufacturers. The following sections—addressing first batch systems then continuous ones—are certainly not all-inclusive. Rather they are intended to provide a good beginning for a filtration search after the filtration testing is completed.

BATCH SYSTEMS

In determining the best batch manner to discharge the cake, there are many varied methods to consider. This section of the guide will provide a succinct overview of several possible systems that might be employed.

Plate and Frame Filter Press

The original filter presses were of plate and frame construction. Modern presses instead use almost exclusively recessed plates. A vast variety of sizes, configurations, plate supports, and degrees of automation exist. There are manual filter presses as well as automatic discharge presses where the plates are hydraulically opened and the cakes fall under their own weight. For sticky cakes, manual intervention is required. Filtration pressures can be up to several thousand pounds per square inch gage (psig).

A new development of the plate and frame filter press is a contained filter press. This press employs a horizontal pressure filter, tubular in construction, with circular plates. The circular plates with welded metal or synthetic media are contained in a pressurized housing. This allows for pressure filtration, cake washing, and vacuum or pressure drying. After the cycle is completed, the housing is moved and automatic cake discharge is via scraper knives that move between the plates.

Horizontal Pressure Leaf Filter

Horizontal pressure leaf filters consist of a number of filter leafs vertically mounted inside the pressure vessel. The filter leaves are normally metal and fitted with a synthetic media. Typically a filter aid (precoat or body feed) is used. Different arrangements exist for removal of the solids, ranging from opening the bottom of the vessel to mechanically spinning or vibrating the cake into a chute. For sticky products, the plates can be sluiced for either cake removal (though this causes dilution) or for leaf cleaning.

Membrane Press

The membrane press is basically a filter press, but instead of having drainage grooves in the plates, the plates are fitted with an inflatable elastomer sheet (which has the drainage grooves). Depending on the operating pressure and product requirements, two different squeezing mediums can be utilized—either air or water. By inflating the sheet at the end of the filter cycle, any residual moisture will be expelled and the cake itself will be squeezed, usually resulting in better cake moisture figures.

Nutsche Filter—Agitated Nutsche Filter

The nutsche or agitated nutsche filters are circular or rectangular filters with a drainage bottom onto which a filter medium is fastened. If the drainage section is connected to a vacuum source, the filters are often open top. If they are closed at the top, they can be pressurized and thus benefit from a higher driving force. This process is based upon thick cakes (normally 2−3 in. up to 12 in. and higher). For this type of filter to be successful, the cake permeability must be able to accept a deep cake without compression.

The nutsche filter may also contain an agitator sealed to the vessel by means of a stuffing box or mechanical seal. The agitator, normally three blades, covers the diameter of the vessel. The agitator moves up and down as well as in clockwise and counterclockwise directions. The agitated nutsche filter can conduct pressure filtration, cake smoothing, cake washing (displacement and reslurry washing), vacuum and pressure drying, and then automatic cake discharge.

Tubular Filter

The tubular filter may be vertical or horizontal and consists of a metal tube and an inflatable membrane. A central filter core consisting of a drainage tube with a filter medium is mounted inside this membrane. The feed is introduced into the tube under pressure. Filtrate exits from the central filter core. After filtration and/or washing the membrane is inflated to squeeze the cake solids to extremely low residual moisture values. To discharge the cake, the membrane is relaxed, the bottom section of the tube opened, and the tube slightly or totally withdrawn to allow the cake solids to drop off or to be scraped off.

Cartridge Filter

Cartridge filters are available in various lengths and diameters as well as construction materials (woven, nonwoven, and membranes).

The flow is outside-to-inside. Therefore, these filters require a strong core to be able to handle the increased pressure differential during operation. Some manufacturers have outer cages for increased stability and capacities. The cartridges are installed in pressure vessels with guide rods for easy installation. The orientation can be vertical or horizontal. The cartridge to vessel seal can be single o-rings, double-o-rings, tie rods, or other designs.

Cartridges are available in cleanable, back-washable metal media or in disposable form. A vast variety of materials, textures, pore sizes, and physical sizes are available. Disposable cartridges can be woven, nonwoven, or pleated and constructed of many types of fibers. The fibers can also be charged to aid in filtration.

High capacity cartridge filters, designed to have increased filter area based upon pleated configuration, flow channels, and flow chambers, are also available. The benefits are increased filter area in small housings when there are space constraints and other process requirements where there are increased solids in the fluid.

Bag Filter

Bag filters, similar to cartridge filters, have various configurations and materials of construction, yet the flow in this process is inside-to-outside. A bag filter normally has a connection for a high-pressure inlet on the top and filtrates exit at the sides and bottom. The solids stay inside the bag. A metal or plastic cage (perforated basket) holds the bag in place during operation. The dirt-holding capacity is the important parameter for the filter design. Depending upon the bag construction—whether it is mesh or felt or single, multilayer, or pleated—the dirt-holding capacity will increase. For instance, a pleated bag has the largest filter area. Bag filters are used for noncritical aqueous applications where the bags and solids can be easily disposed.

Candle Filter

Candle filters provide for thin-cake pressure filtration, cake washing, drying, reslurry, and automatic discharge as well as heel filtration in an enclosed pressure vessel. Units are available from $0.17\,m^2$ up to $200\,m^2$ of filter area per vessel.

Figure 3.1 Three components of filter candles.

Filter candles (Figure 3.1) consist of three components: single-piece dip pipe for filtrates and gas, perforated core with outer support tie rods, and filter sock.

The filtrate pipe is the full length of the candle and ensures high liquid flow as well as maximum distribution of the gas during cake discharge. The candle can be a synthetic, stainless steel, or higher alloys. The outer support tie rods provide for an annular space between the media and the core for a low-pressure drop operation and efficient gas expansion of the filter media sock for cake discharge (Figure 3.2).

The filter media is synthetic with a clean removal efficiency to <1 micron (μm). As the cake builds up, removal efficiencies improve to <0.5 μm.

The candle filter vessel (Figure 3.3) is constructed of stainless steel or higher alloys. Within the vessel are horizontal manifolds called candle registers. Each candle is connected to a register with a positive seal to prevent bypass. Each register may contain from 1–20 candles

Figure 3.2 Outer support tie rods provide annular space.

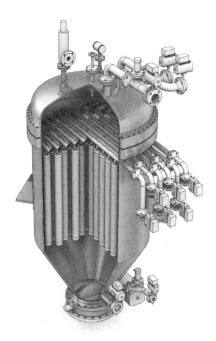

Figure 3.3 Candle filter vessel construction.

depending upon the filter size. The registers convey the liquid filtrate in the forward direction as well as the pressure gas in the reverse direction for filter media sock expansion. Each register is controlled with automated valves to ensure optimum flow in both directions.

Pressure Plate Filter
There are three main types of plate filters based upon cake discharge: manual, vibrating plate, and spinning plate.

The designs consist of horizontally arranged filter plates mounted together to form a set of plates or plate stack. The filtrates will be discharged upward through the hollow shaft. The set of plates is fixed to the hollow shaft by means of four tie rods or other designs. A backing screen, of metal or synthetic fabric and the filter cloth, is mounted on the filter plates. The filter cloth is fixed and sealed at the periphery of the plates by means of tension rings or welding. A stuffing box, pressurized stuffing box, or mechanical seal can be used to seal the hollow shaft to the vessel.

Cake removal varies across the plate filters. In the manual discharge, the plates are lifted out of the vessel and the cake is manually scraped off the plates. In the vibratory design, two unbalanced motors gently vibrate the plates to cause the cake to move in an elliptical fashion as well as up to down to come off the plates. Gas-assist (reverse flow of gas down the filtrate shaft) is possible to lift the cake off the plate before vibration. In the spinning plate design, there are drive motors, gear box assemblies (including gear box housing, gear reducers, bearings, shaft bearing arrangement and bushings, torque loadings, etc.), and mechanical seals (either single or double with special maintenance and cleaning) allowing the plates to spin. With spinning plates, there are some unique installation challenges such as center of gravity, static and dynamic balancing, and bearing lubrication.

Combination Filter Press-Membrane Press
This horizontal pressure filter is tubular in construction and operates as a contained filter press. Circular plates with welded metal or synthetic media are contained in a pressurized housing. During feeding, the elastic membranes uniformly join together with lowest tension at the core plate. The filter cake thickness after squeezing is less than the chamber depth, allowing for pressure filtration, cakes washing, and

vacuum or pressure drying. After the cycle is completed, the housing is moved and cake is automatically discharged via scraper knives moving between the plates.

Tower Press

The tower press utilizes vertically stacked chambers to dewater slurry and form a filter cake. The tower press has horizontal square plates with recessed membrane chambers. Each chamber can be isolated for individual operation. The cloth zigzags through the plate stack and is moved at the end of each filter cycle to completely discharge the filter cakes. High-pressure water sprays wash the filter cloth as it leaves the plate stack. The unit is operated by hydraulic pressure up to several thousand psig.

CONTINUOUS SYSTEMS

A second approach to filtration utilizes continuous systems. This section introduces some among the range of options available to the process engineer considering this methodology.

Rotary Pressure Filter

The rotary pressure filter is a continuously operating unit for pressure filtration, cake washing, and drying of slurries up to 50% solids. The filter houses a rotating drum with cell inserts fitted within the filter media. Filtration is conducted via pressure of up to 6 bar. Positive displacement washing or countercurrent washing follows filtration. Multiple washing steps as well as solvent exchanges, steaming, and extraction are accomplished. Finally, the cake is dried by means of blowing hot or ambient gas.

This filter has a uniquely designed discharge system providing for atmospheric discharge from pressure filtration. After automatic cake discharge, the filter cloth is washed; the clean filter cloth then re-enters the feeding/filtration zone thereby continuing the process. All solvent and gas streams can be recovered separately and reused in the process to minimize consumption. The rotary pressure filter is used for applications requiring maximum washing efficiency and containment. The overall cycle time for filtration, cake washing, and drying is <2 min.

Disc Filter

These consist of several flat discs, up to 15 in., each made up of sectors clamped together to form a disc. The discs are mounted on a hollow shaft and ribbed toward the neck of the filter. Each disc is made of metal and has on either side an open cloth support structure connecting with the hollow shaft that carries the filtrate. A filter cloth is fastened to the disc and the hollow shaft connects to a vacuum source. The disc, usually 30% submerged, rotates slowly in a feed trough where it picks up the solids. These can later be scraped off just prior to re-entry into the feed trough. The disc filter has applications for fast filtering crystals with few washing or drying requirements.

Rotary Drum Vacuum Filter

The original rotary drum filter consisted of a cylindrical drum, fitted with a drainage grid on the outside over which a filter cloth was stretched. The filter was supported by two bearing blocks and composed of successive cells, which formed a horizontal cylinder. The drum had perforations and was connected to a vacuum source. The drum itself was mounted on a horizontal shaft and submerged about 30% in a feed trough. The slurry passed through the trough-shaped tank to be filtered. The tank, part of the filter frame, also supported the drum, the cake discharge system, and the slurry agitation device. The vacuum lifted the slurry out of the trough and caused the liquor phase to be sucked through the cloth, leaving the solids behind as a filter cake. The cake was scraped off just before re-entering the feed trough. The newer designs have additional cake washing and drying steps as the drum rotates. The washing is on the front side of the rotation; if this was on the back side, the cake may fall off. The drying is on the back side of the rotation. In addition, the newer designs can use a filter aid for precoat where a controlled scraping off of the cake (by layers) is possible.

Pressurized Drum Filter (P-DF)

The pressurized drum filter (P-DF) is a rotating drum inside a pressure vessel. The unit consists of a filter drum, slurry trough, agitator, wash bars, and a pressure let-down rotary valve. The process begins by closing the pressure vessel manually with bolts and nuts and pressurizing the vessel with compressed gas. The rotary valve is also pressurized for sealing and the filter trough is filled via the suspension feed pipe. The agitator is then started to keep the solids in suspension. Cake washing and gas drying by blowing follow. The filter media is inflated by a gas blow to discharge the cake. A scraper-deflector plate directs the cake into the chute. The chute is connected to the rotary valve for continuous cake discharge. The P-DF has applications with deep cakes of up to 6 in. and large filter areas; however, as the area becomes larger, there is a larger amount of gas consumption.

Vacuum Belt Filters
The two main types of vacuum belt filters are carrier belt and tray (fixed or indexing). A continuous belt subjected to a vacuum source is the filter medium. Feed introduced at one end is normally allowed to settle under gravity for a few seconds. This causes the coarser fraction to form its own precoat before the suspension is subjected to full vacuum. Controlling the cake's thickness by adjusting speed of travel or indexing time achieves the optimum cake thickness to obtain the required residual moisture and/or cake washing efficiency. The horizontal configuration makes this filter one of the most efficient cake washing filters. Further, the vacuum belt filter can have many washing stages including countercurrent washing.

Pan Filter
In its basic form, the pan filter is a filter tray consisting of a filtrate collection box connected to a vacuum source. A drainage grid, suitably clothed with a filter cloth, is mounted inside the box. The whole arrangement can be tipped over 180°, causing the cake to drop out. This obviously is a batch filter but the same principle is used in the rotary pan filter, where a very large wheel of several meters in diameter is formed through triangular filter trays connected to a central drive and tipping mechanism. The wheel rotates the filter segments through the feed, dewatering, cake wash, and final drying stages after which each individual tray is turned 180°. Cloth washing sometimes follows. The filtration mechanism and efficiency is very similar to that of a vacuum belt filter. Although very large filter areas can be offered, the circular shape nets very large floor space considerations.

Screw Press
The horizontal continuous screw press contains a conveying screw rotating inside a perforated cylinder. The incoming slurry is forced the length of the machine, and liquid (press liquor) is expelled through the cylinder wall. The remaining solids (press cake) are discharged at

the far end. The screw has a graduated pitch and interacts with stationary resistor teeth. These teeth assure that the material to be pressed (i) does not turn (co-rotate) with the screw and (ii) is intermixed during passage through the press. As a result, more material is exposed to the screening surface. The graduated pitch screw gradually compresses the material as it passes through the press's main screen cylinder. The press liquor is forced through the perforations in the screen. The press cake encounters resistance at the cylinder discharge in the form of a cone mounted on a ram. As the material reaches the discharge point, the cone exerts final pressing action to achieve maximum dewatering. The cone system allows easy adjustment of the resistance to cake discharge.

Belt Press

Similar to a screw press, the belt press applies mechanical pressure to a chemically conditioned slurry sandwiched between two tensioned belts as the belts pass through a serpentine of decreasing diameter rolls. The machine can actually be divided into three zones: gravity zone, where free draining water is drained by gravity through a porous belt; wedge zone, where the solids are prepared for pressure application; and pressure zone, where medium, then high pressure is applied to the conditioned solids. Some belt presses are available in three-belt or an extended gravity two-belt design. The three-belt press has an independent gravity zone with a more open belt for more rapid drainage of the volume of water. The extended gravity design has a longer gravity drainage zone.

Having provided a broad overview of the various approaches available to the process engineer within the batch and continuous filtration systems, this guide turns next to a consideration of combination filtration.

CHAPTER 4

Combination Filtration

Filtration experts have long debated the definition of combination filtration. In the realm of cartridge filtration, simply defined, a combination filter is one that does at least one other processing job at the same time as filtering a suspension. For example, this could be a carbon canister that removes both suspended and dissolved components. In water applications, a combination filter removes bacteria, sediment, chlorine taste and odor, and scale. In lubrication oil filtration, combination filtration refers to full-flow and bypass flow filtration. Finally, for small-scale process filtration, combination filtration is installing bag and cartridge filtration systems in series.

There is, however, a new definition of combination filtration that transcends the standard approach and will assist process engineers with troubleshooting and idea generation. The approach relies upon the slurry analysis and testing to uncover the process symptom. A process solution called combination mechanical slurry conditioning and filtration is then developed.

There are, without doubt, many technologies that can be applied in combination, including the use of chemicals such as flocculants and coagulants. However, from a practical viewpoint, this chapter focuses on a brief description of examples of different types of filtration technologies that can be combined based upon general operating conditions at chemical plants. Combining technologies does increase the capital cost. Yet, in decision making, it is important to note the overall reliability and operating costs of the project will simultaneously be optimized for maximum efficiency.

TESTING FOR COMBINATION FILTRATION

Combination filtration testing (Figure 4.1) builds upon the initial information on filtration testing discussed in Chapter 2. Still, additional data is needed. An important parameter for concentrating filtration is

---

1. Test purposes
The purpose of these tests is to:
1.1 Determine required filter area for the primary filtration.
1.2 Evaluate filtrate quality of both primary and secondary filtration.
1.3 Determine filter aid requirement for secondary filtration.
1.4 Evaluate washing efficiency and minimize solvent loss in filter cake.

2. Test methods
The leaf filter, jacketed with a 400 ml capacity and with 20 cm^2 filter area, can be used to gather data and make observations on this product.

The following information was gathered during this test:

Primary filtration test methods:
2.1 Variable speed peristaltic pump was used to conduct primary filtration tests.
2.2 Filtration rate versus differential pressure
2.3 Filtrate quality versus filter media

Secondary filtration test methods:
2.4 Filtration pressure for secondary filtration tests was provided by compressed air.
2.5 Filtration rate versus filter-aid concentration
2.6 Filtrate quality versus filter-aid concentration
2.7 Filtration rate versus filter media
2.8 Filtrate quality versus filter media
2.9 Filter cake residual solvent or filtrate conductivity versus volume wash water
2.10 Filter cake residual moisture versus drying air usage

Figure 4.1 Testing for combination filtration.

the settling time of the solids after the initial filtration. Settling is impacted by the solid's density, liquid density, quantity and nature of the solids, and more. For testing, it is important to perform batch settling tests at different solid concentrations to see the variations in the terminal settling velocity.

THICKENING

There are many different ways to employ combination filtration. The following section considers several types and variations of thickening/concentrating applications in which the goal was met using a combination of processes.

Concentrating Candle Filters Followed by Batch Pressure Plate Filtration

In this application, the solids consist of catalyst fines with a flow rate of 260 gpm with an overall concentration of 600 ppm. The solids are very fine and the specification required removal of 99% particles >1 micron.

Figure 4.2 Concentrating candle filters followed by pressure plate batch filtration.

Two further requirements are to minimize solvent losses in the cake (i.e., maximum dryness) and to have the lowest operating costs possible such that compressed air usage and di-ionized (DI) water usage are minimized. Another constraint for the DI water is the plant capacity in addition to high use of DI water would dilute the process. The initial idea is the use of candle filters for the entire clarification process. The testing results in the requirement of 80 m^2 of filter area. While that area is not overly large, the DI water and compressed gas usage would be very high. As a result, a two-stage process is presented (Figure 4.2).

In the first stage, the suspension was filtered/concentrated in the candle filters without filter aids or chemicals, and prethickened on a continuous basis. In this approach, similar to the process in settling tanks, a solid's concentration is achieved that is capable of being further processed. These filters are able to retain particles down to 0.5 micron and discharge these solids as a concentrated sludge/slurry. Depending on the solid's content, the time for concentrating is between 2 and 8 h. In order to reduce utilities, no drying or washing is necessary in the larger candle filters.

As soon as a filter cake of a few millimeters has built up through filtration, the suspension feed is stopped and the cake is displaced from the candles. The cake settles for approximately 30 min and collects in the cone of the filter vessel. The concentrated slurry that is discharged has a solids content of around 3−6% from an initial concentration of <1%.

This material is pumped into an agitated vessel. The prethickened suspension is then transferred to a batch pressure plate filter. The concentrated slurry will then be filtered on a secondary pressure plate filter to recover the remaining solvent and discharge the dry solids. In the pressure plate filter, the circular plates are arranged in a horizontal plate stack. This results in a stable cake for washing and drying without the risk of the cake sliding off.

The secondary filtration is a batch process incorporating filtration, countercurrent washing, drying, and dry cake discharge. A residual solvent content of $<5\%$ is achieved with a low wash ratio. Using closed cycle washing or countercurrent washing minimizes water consumption. The final crumbly and low volume product can be disposed of as a nonhazardous cake as there is only a small trace of solvents. The overall result is a very dependable process with high quality filtrate and minimal utility air and water usage, a 40-fold reduction.

Concentrating Candle Filters Followed by Continuous Vacuum Filtration

In this application, the slurry feed comes from an upstream candle filter thickener. The primary goal is to dry the waste solids for landfill. The secondary goal is to filter the liquid stream. Tests indicate the polymer/salt waste product cannot be dried to more than 65% liquid (by weight) before the cake cracks. The dried separated solid drops by gravity flow to a screw conveyor.

The initial slurry feed is in the range of 2% solids. For a vacuum belt filter, this is very low solids such that the belt filter sizing is based upon hydraulic liquid loading rather than dry solids. The result is a large required filter area. With the use of concentrating candle filters, the percentage of solids in the slurry feed can be increased to 12% and a significant reduction in the filter area—50% of the area required based upon hydraulic loading. More importantly, a smaller filter area requires a smaller liquid-ring vacuum pump and the associated energy savings.

In this case (Figure 4.3), the candle filters operate to mechanically condition the slurry through thickening and concentrating up to 12% solids. The resulting slurry can then be economically and technically processed on a vacuum belt filter by vacuum filtration, cake washing, and drying. The drying can be by vacuum, compression, blowing with hot or ambient-temperature gas, or steaming. Applications for this approach occur in chemical plants as well as in coal gasification plants.

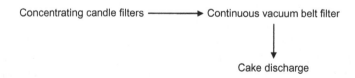

Figure 4.3 Concentrating candle filters followed by continuous vacuum filtration.

Candle Filtration Followed by Conventional Filter Press

In some applications, a conventional filter press is the correct technology for dewatering but the initial slurry feed is not optimal in terms of the percent solids as well as the size of the particles. For these applications, mechanically conditioning the slurry by concentrating candle filters is the optimum approach. The concentrating candle filters will increase the solids concentration in the slurry as well as reduce the volume of slurry that is to be treated by the filter press. The candle filters will also produce a high quality filtrate (effluent) such as total suspended solids (TSS) <1 ppm with a particle size of <0.5 micron (Figure 4.4). Specialty chemical, chlor-alkali, and coal gasification plants are typical examples. For chlor-alkali plants, concentrating candle filters are used upstream of the brine evaporator to remove solids and then downstream. The concentrated sludge is then sent to the filter presses for final dewatering (Figure 4.5).

Figure 4.4 Concentrating candle filters followed by conventional filter press filtration.

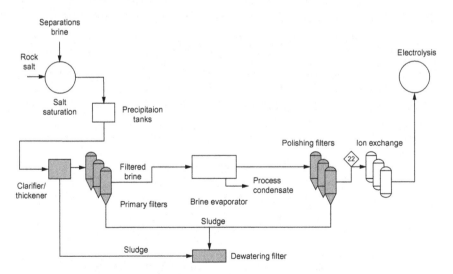

Figure 4.5 Chlor-alkali brine filtration with concentrating candle filters followed by conventional filter press filtration.

POLISHING

Combination filtration can also have applications in polishing processes downstream of pressure and vacuum bulk separation technologies. The following applications illustrate some of the most effective methodologies.

Continuous Vacuum Filtration Followed by Concentrating Candle Filtration

In this case, the specialty chemical slurry, used as a flame retardant, contains a high solids loading that is optimum for a vacuum belt filter but with a very small particle size distribution resulting in slow filtration. The solids are also very sticky. The process objectives are for cake washing to a very low conductivity with final moisture content to be <35% and clear filtrates of <5 g/l. While continuous vacuum filtration solves the first part of the filtration process (cake washing and drying with a cake thickness of 5−6 mm), the filtrate clarity is not achieved and the sticky cake results in difficult discharge.

The question facing the process team is how to optimize the filtration rate, to improve the clarity of the mother filtrate containing small, very fine solids, and to improve the cake discharge to maximize the production yields.

The decision is to use continuous vacuum filtration followed by concentrating candle filters (Figure 4.6). First, a more open filter cloth improves filtration rates and cake discharge. While the more open cloth allows more solids to pass, concentrating candle filters removes and recovers these solids from the mother filtrate, concentrates them, and then sends them back to the process. Secondly, as the solids are sticky, the cloth wash contains solids. These solids, as above, are also filtered and concentrated in candle filters and then sent back to the process.

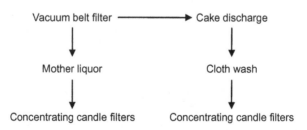

Figure 4.6 Continuous vacuum filtration followed by candle filtration.

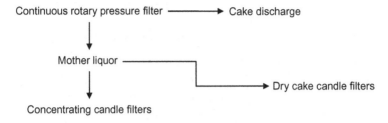

Figure 4.7 Continuous rotary pressure filtration followed by candle filtration.

Continuous Rotary Pressure Filtration Followed by Candle Filtration

For this commodity chemical, with rates up to 60 tons of dry solids/ hour, the initial mother liquor contains fines in the 1- to 5-micron size range. These needle-like fines make their way through a deep, 6-in. cake into the mother liquor. The process decision is to accept fines in the mother liquor rather than have a reduced rate with a tighter filter cloth. The fines are then recovered in the candle filters as they can either produce a dry cake or concentrated slurry. If it is a concentrated slurry, it will be sent back to the reactor for a dry cake discharge. The cake would be added into the dry cake from the rotary pressure filter (Figure 4.7).

PROCESS SEGMENTATION

Engineers may also consider the problem-solving value of combination filtration in process segmentation. For example, rather than attempting to find one technology for the process, it may be better to separate the process into different steps. The following applications help illustrate the value of mixed methodologies.

Continuous Vacuum Filtration Followed by Contained Filter Press Filtration

In this pharmaceutical process, there are three unique process reaction steps occurring that are being processed by only two types of filtration technologies (Figure 4.8). The first reaction requires vacuum filtration, cake washing, and drying by vacuum only as the crystals are fragile and cannot be mechanically pressed. The continuous-indexing vacuum belt filter can have the pressing device installed but not operated. The second step has more robust crystals and requires a lower moisture specification. The cake from the second reaction is then processed on the

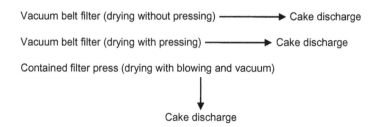

Figure 4.8 Continuous vacuum filtration (two variations) followed by contained filter press.

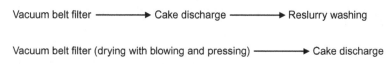

Figure 4.9 Continuous vacuum filtration (two variations).

same vacuum belt filter but, in this case, the cake is mechanically compressed with blowing for final drying. The cake from the second step then undergoes further processing, but, in this case, pressure filtration is the optimum solution. Due to the nature of the chemicals and solvents, a contained filter press for pressure filtration, cake washing and drying, by both blowing and vacuum is accomplished.

Continuous Vacuum Filtration with Reslurry Washing—Multiple Uses

In this case, similar to a previous example, it is deemed more efficient to use the same vacuum belt filter with two different variations. While filtration and drying are easily accomplished, the washing step is the rate-limiting step. If the vacuum belt filter is operated at two-thirds of the production rate, causing a production bottleneck, adequate washing is still not possible. This is in spite of a longer residence time due to more zones. The most likely cause of the performance difference is the slurry and the particle size distribution and shape. After filtration, the particles may change shape, form agglomerates, and compress, among other attributes. While the particle size distribution looks the same, there is also the possibility of other variable forces impacting the cake.

The decision is to employ a dual process (Figure 4.9). First, a vacuum belt filter is used in the process to remove the initial mother liquor with no washing or drying. The cake is then slurry discharged from the vacuum belt filter back to an agitated slurry tank where dilution water washes the cake to achieve the desired quality. The reslurried cake is then fed back to the vacuum belt filter to be washed and dried at a high throughput. This may require less overall water given the initial removal and enable faster processing speeds for the filter.

Throughout, this chapter has discussed a new definition of combination filtration providing process engineers a framework for idea generation when analyzing an operating bottleneck. The approach relies upon the slurry analysis and testing to uncover the process symptom and then develop a process solution called combination mechanical slurry conditioning and filtration. This chapter's applications illustrate that installation and combined use of filtration technologies, while higher in capital cost, will result in a more reliable operating process at the plant. This guide next turns its attention to filter selection.

CHAPTER 5

Filtration Selection

Filtration selection, harkening back to Sherlock Holmes, requires "not jumping to conclusions." There is no "one size fits all" process solution. Selecting a filtration technology requires a systems approach that must be incorporated with other solids processing such as reactors, dryers, solids handling, and others.

For example, in a typical bioseparation process, downstream of the bioreactor, there are filtration choices to be made depending upon the type of product (intracellular or extracellular) and how to remove the biomass. The selection process includes vacuum filtration, centrifugation, pressure (press) filtration, candle filtration, or a simple flotation process (Harrison, 2014).

In this respect, Coulson and Richardson's familiar *Chemical Engineering Design* summarizes the approach very well stating that the engineer "is not normally involved in the detailed design of the equipment. The engineer's job is to select and specify the equipment needed for a particular process while consulting with the vendors to ensure that the equipment is suitable" (Jacob & Collins, 2014).

The process has three components that must be considered: material properties, mechanical properties, and separation performance. These are combined and the ranked choices must then be evaluated weighing operational, economic, and plant (internal and external) objectives.

The material properties examine the solids and the liquids. For solids, the engineer will need to know the total suspended solids (TSS) and solids concentration, particle size distribution (PSD), and particle shape. The PSD should be based upon particle counts at different sizes rather than by weight or volume as this will provide the equipment vendor the most accurate information.

Particle shapes can be spheres, rounded, angular, flaky, or thinly flaked, among others. These shapes will influence the filtration rates

Table 5.1 Impact of Particle Size on Process		
Technique	Basic Measurement	Particle Variables Affecting Measurement
Laser light scattering	Light intensity versus scattering angle	Refractive index, shape, orientation
Electrical sensing zone	Change in electrical signal across a conducting orifice	Porosity, conductivity
Sedimentation	Settling velocity	Density, shape
Dynamic image analysis	Linear dimensions of projected cross-section of particle	Particle orientation

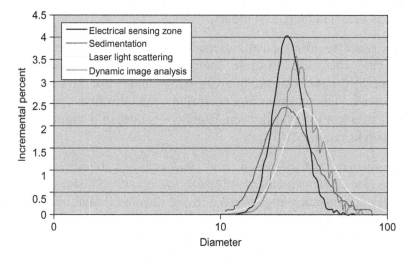

Figure 5.1 Testing for garnet.

for the process and also impact the PSD due to the nature of particle size measuring equipment (Table 5.1).

To further illustrate this point in practical terms, consider the variations in testing results for three different types of particles: garnet (irregularly shaped silicate minerals), wollastonite (smooth calcium inosilicate minerals), and glass spheres (perfect spheres).

Garnet crystals approximate cubic shapes. Since the diagonal of a cube is approximately 30% longer than a sphere of the same volume, larger particle size is reported by methods that take orientation into account (laser light scattering and dynamic image analysis) (Figure 5.1).

On the other hand, wollastonite particles are rod shaped. Since the apparent dimensions of a rod-shaped particle can vary drastically

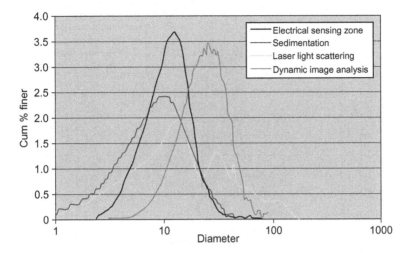

Figure 5.2 Testing for wollastonite.

depending on its orientation, the detection methods that were affected by particle orientation (laser light scattering and dynamic image analysis) exhibited larger particle size measurements and broader peaks (Figure 5.2).

Glass spheres, as expected, produced the most consistent results for the different techniques. Since the particles are spherical, their orientation has no effect on their measurement. Microscopy of the sample indicated that the glass spheres contained air bubbles in variable sizes, reducing somewhat the density of some of the spheres. Because of this, the device detects some of the particles as being undersized and, therefore, widens the distribution and shifts it to a slightly finer size (Figure 5.3).

For the liquids, the typical parameters include viscosity, temperature, and the relationship of the two, vapor pressures, pH, ionic strength, and any other unique conditions.

In terms of mechanical components, the engineer will need to provide information about materials of construction, temperature, pressures, seal information, compatibilities of the solids, liquids, cleaning solutions, and other characteristics. While this may appear simple, consider the example of the selection of o-rings. This is just component, yet an o-ring can be EPDM (rubber) and there are many different types of EPDM such as natural rubber, peroxide-cured, etc. Careful selection of each component is necessary. Other questions can include,

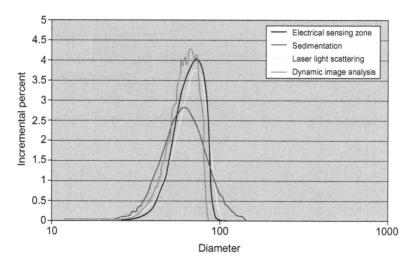

Figure 5.3 Microscopy of glass spheres.

but are not limited to: What is the level of containment? Are there space constraints at the plant? Is it a complex process?

Finally, there is the separation performance, which includes capacities or throughputs of solids, liquids or dry solids, filtrate quality, dryness, washing liquids, conductivity, wash ratios, and more. Is the product the liquid or the solid or both? The more information the engineer can provide about the process and the requirements, the better the accuracy of the vendor's information. The following section discusses mechanical and process questions that lead to the required specifications for the equipment vendor.

SPECIFICATIONS

Each equipment vendor as well as engineering companies will have different and unique process questionnaires or specification checklists to employ. Tables 5.2–5.4 are typical examples that can be tailored to meet the objectives of the process. These examples will provide a baseline for review.

In determining the process specifications, the engineer must also consider the upstream and downstream equipment as discussed later.

Table 5.2 Mechanical Properties Checklist

1. Design
2. Filter type
3. Piping
4. Steel structure or skid package
5. Controls
6. Valves
7. Instruments
8. Shipping
9. Specifications
10. Codes and standards
11. Documentation
12. Customer specific requirements

Table 5.3 New Filtration Process Questionnaire

1. Customer's data

Company: _____   Date: _____

Contact person: _____   Department: _____

Address: _____   Phone: _____

City, ZIP: _____   Fax: _____

State: _____   E-mail: _____

2. Process data

2.1 Short description of the new process (include sketch or data sheet if necessary)

(name of the process, stages upstream and downstream of the filtration or separation):

_____

_____

_____

_____

_____

_____

_____

_____

_____

_____

_____

_____

Table 5.3 (Continued)

2.2 Process Conditions:

Solids throughput (per filter):	Minimum _____ t/h	
	Nominal _____ t/h	
	Maximum _____ t/h	
Process:	☐ Continuous	
	☐ Discontinuous, batch duration: _____ min	
	batch volume: _____ t	
Moisture content:	Nominal _____ wt-%	
	Minimum _____ wt-%	
Cake characteristic:	☐ Thixotrope ☐ Abrasive ☐ Sticky	
Flocculants used:	☐ No	
	☐ Yes, dosage: _____ g/m³	
	Type: _____	
Filter aid used:	☐ No	
	☐ Yes, dosage: _____ g/m³	
	Type: _____	
Usage of flocculants or filter aid:	☐ Allowed ☐ Not allowed	
Explosion-proof:	☐ Yes ☐ No	

3. Which benefit of the filtration process may be of interest for you?

	Very important	Less important	Not important
- Moisture content	☐	☐	☐
- Cake washing	☐	☐	☐
- Avoidance of particle breakdown	☐	☐	☐
- Low solids in the filtrate	☐	☐	☐
- Low space requirement	☐	☐	☐
- Others: _____			

_____
_____
_____
_____

4. Components

	Product	Waste/Disposal
- Solids / cake	☐	☐
- Mother liquor	☐	☐
- Wash filtrate	☐	☐
- _____	☐	☐

Additional remarks: _____
_____

Table 5.3 (Continued)

---

5. **Suspension data (filter feed)**

Solids concentration: c =   _____ g/l or   _____ wt.% or   _____ vol.%

Temperature: Normal: _____ ° C   Maximum: _____ ° C

5.1 **Suspension liquor / filtrate**

Name: _____

Chemical composition: _____

Liquid density: _____ g/cm^3 (at T = _____ ° C)

pH value: _____

Max. allowed solids content in the filtrate: _____ g/l

Additional remarks (e.g., viscosity): _____

5.2 **Solids material / filter cake / concentrate**

Name: _____

Chemical / mineralogical composition: _____

Solids density: _____ g/cm^3 (at T = _____ ° C)

Particle size: x_{10} = _____ μm x_{50} = _____ μm x_{80} = _____ μm

Particle shape: _____

Particle characteristics: ☐ Gel-like   ☐ Amorphous

                                  ☐ Crystalline   ☐ Fibrous

Decomposition temperature: T = _____ ° C

Additional remarks (e.g., ash content, solubility): _____

5.3 **Washing procedure (if applicable)**

Washing liquor   ☐ Temperature: _____ ° C

                            ☐ Name: _____

Washing liquor availability ☐ Total (per filter)   _____ m³/h

Wash efficiency   ☐ Parameter controlled _____

                            ☐ Start value (and unit) _____

                            ☐ End value (and unit) _____

Washing procedure / washing liquor distribution (short description):

_____

_____

---

6. **Handling of suspension / solids / liquid (for lab tests)**

Any special instructions for handling?   ☐ Yes ☐ No

Will safety data sheet be handed over?   ☐ Yes ☐ No

Is the suspension aging?   ☐ Yes ☐ No

Maximum storage time without changes in product characteristic (e.g., x_{50})? _____ days

Additional remarks: _____

_____

_____

Table 5.3 (Continued)

7. Questions / additional remarks

(e.g., experiences with lab filters, flocculants, features, desired time schedule, etc.):

_____

_____

_____

_____

_____

_____

Table 5.4 Existing Filtration Process Questionnaire

1. Customer's data

Company: _____   Date: _____

Contact person: _____   Department: _____

Address: _____   Phone: _____

City, ZIP: _____   Fax: _____

State: _____   E-mail: _____

2. Process data

2.1 Short description of the current process (include sketch if necessary)

(name of the process, stages upstream and downstream of the filtration or separation):

_____

_____

_____

_____

2.2 Filters currently installed:

- Make / manufacturer / type of filter: _____
- Filtration area (m²) per filter: _____
- Number of filters installed: _____
- Year of commissioning: _____
- Additional remarks: _____
- Filter medium: Type / material: _____ Mesh size: _____ μm

    Manufacturer: _____   Name: _____

Table 5.4 (Continued)

2.3 Current filtration and process conditions:

Solids throughput (per filter):	Current _____ t/h _____ kg/m²h
	Aimed _____ t/h _____ kg/m²h
Process:	☐ Continuous
	☐ Discontinuous, batch duration: _____ min
Moisture content:	Current _____ %
	Aimed _____ %
Pressure difference:	Δp = _____ bar _____ kPa _____ mmHg
Cake thickness:	_____ mm
Filter speed:	_____ rpm
Cake characteristic:	☐ Thixotrope ☐ Abrasive ☐ Sticky
Flocculants:	☐ No
	☐ Yes, Dosage: _____ g/m³
	Type: _____
Filter aid:	☐ No
	☐ Yes, Dosage: _____ g/m³
	Type: _____
Usage of flocculants or filter aid:	☐ Allowed ☐ Not allowed
Explosion-proof:	☐ Yes ☐ No

3. Which benefit of the filtration process may be of interest for you?

	Very important	Less important	Not important
- Improved solids throughput	☐	☐	☐
- Lower moisture content	☐	☐	☐
- Improved cake washing	☐	☐	☐
- Avoidance of particle breakdown	☐	☐	☐
- Less solids in the filtrate	☐	☐	☐
- Automated cake discharge	☐	☐	☐

4. Separation objective

_____

_____

_____

Table 5.4 (Continued)

5. **Suspension data (filter feed)**

Solids concentration: c =   _____ g/l   _____ wt.%   _____ vol.%

Volume throughput: Current (per filter):   _____ m³/h   _____ m³/m²h

Aimed (per filter):   _____ m³/h   _____ m³/m²h

Temperature: Normal: _____ °C   Maximum: _____ °C

5.1 Suspension liquor / filtrate

Name: _____

Chemical composition: _____

Liquid density: _____ g/cm³   (at T = _____ °C)

pH value: _____

Max. allowed solids content in the filtrate: _____ g/l

Additional remarks (e.g., viscosity): _____

5.2 Solids material / filter cake / concentrate

Name: _____

Chemical / mineralogical composition: _____

Solids density: _____ g/cm³   (at T = _____ ° C)

Particle size: x_{10} = _____ µm x_{50} = _____ µm x_{80} = _____ µm

Particle shape: _____

Particle characteristics: ☐ Gel-like   ☐ Amorphous

☐ Crystalline   ☐ Fibrous

Decomposition temperature: T = _____ ° C

Additional remarks: _____

5.3 Washing procedure

Washing liquor   ▪ Temperature: _____ ° C

▪ Type: _____

Washing liquor consumption   ▪ Total (per filter):   _____ m³/h

▪ Specific: _____ m³/t

Washing liquor distribution / washing procedure (short description):

_____

_____

_____

Table 5.4 (Continued)

6. **Handling of suspension / solids / liquid**

   Any special instructions for handling?   ☐ Yes ☐ No

   Will safety data sheet be handed over?   ☐ Yes ☐ No

   Is the suspension aging?   ☐ Yes ☐ No

   Maximum storage time without changes in product characteristics___hours / days

   Additional remarks: _____

   _____

   _____

7. **Questions / additional remarks**

   (e.g., experiences with lab filters, flocculants, features, desired time schedule, etc.):

   _____

   _____

   _____

   _____

UPSTREAM AND DOWNSTREAM EQUIPMENT

Always remember the solid-liquid filtration system requires a systems approach that must be incorporated with other solids processing such as reactors, dryers, and solids handling. The scope of this guide is not to discuss the actual upstream and downstream equipment but rather to remind engineers that these other components must be part of the filtration equipment discussion.

The following is a typical example of a chemical process that includes all of the associated processing steps:

- Chemical synthesis and crystallization
     Types of catalysts
     Solvents
     Continuous or batch
     Temperature
     Flashing
- Inerting
- Slurry handling
- Filtration
- Drying

- Dissolution
- Hydrogenation
- Secondary crystallization
- Filtration
- Final drying
- Solids handling

From the above list, slurry handling and solids handling are critical parameters as they impact the operation of the filtration system. In terms of slurry handling, each equipment vendor will have a recommendation about the type of pump required, feed pressures at the filter, and more. The piping to the filter feed, for example, can also impact the filtration operation if there is settling, increased solids velocity, dynamic versus static pressures, or other factors.

Finally, the solids discharge from the filtration system and solids handling to the downstream equipment can result in a bottleneck if not carefully watched. Each type of powder behaves differently and sometimes counterintuitively. It is for this reason that a comprehensive understanding of the properties of each powder is necessary including the impacts of moisture, angle of repose for flow, bulk density and bulk permeability, shear stress, dynamic flow, and other properties (Armstrong, Brockbank, & Clayton, 2014).

Having given consideration to equipment, the engineer must also then examine integration and controls.

INTEGRATION AND CONTROLS

Continuing with the project, now that the process filtration selection has been made and the mechanical specification (user requirements) has been written taking into account the upstream and downstream equipment, the next step is to look at the process integration and process controls. This can be very difficult for the process engineer as control requirements involve other groups and, in reality, electrical/controls engineers speak an entirely different language compared with chemical engineers.

The first discussions will need to be with the process filtration supplier to obtain their recommendations for how to control the system. There are normally three choices: manual operation to be controlled by the operators, local programmable logic controller (PLC), or a distributed control system (DCS). The complexity of the unit as well as the upstream and downstream process and equipment will begin to determine the correct control scheme.

Normally, there are two documents required for the integration. Remember that once these documents are completed, the controls integration portion of the project runs parallel to the mechanical portion of the project with drawing approvals, etc.

The first document will be from the operating company and it will define the overall controls philosophy and standards. The typical sections include:

Purpose: Provides the functional specifications for the automation and controls associated with the computer interaction for the process technology.

Scope: Describes the control system functional and operational specifications for how the filtration system and any other upstream and downstream process equipment will interface with each other. This will cover the control panels, human–machine interface (HMI), emergency and safety interlocks, process interlocks and indications, maintenance, manual, and automatic controls.

References: Describes the specific industry or other standard documents such as the ISA S88 Standard—ANSI/ISA-S88.01-1995: Batch Control and NEMA, NEC and NFPA documents.

Definitions/acronyms: This is an important section such that the supplier and operating company will agree on terminology.

Functional requirements: Covers the purpose of the system, expected results, relation to the other process systems, controlled functions, interfaces, information flow, material flow, and finally cycle descriptions (in flow chart form).

Information: Outlines analog inputs, discrete inputs, operator and engineering inputs and set points (including password protections, as necessary), recipes and process information, calculations, outputs, alarms, start-up/shutdowns, emergency operations, displays, reports, alarm management, safety, security, and failures.

Physical: Reports system structure, layouts, relation to other systems, control hardware, HMI stations, data storage and historical trending, configurations, display colors and design, communication and networking, redundancy/reliability/dependability, materials/utilities/environmental, and electrical interfaces.

The second document required for integration will be from the supplier. This will match their specific designs to the operating company document above. The typical supplier functional description specification (FDS) should include:

1. General
   1.1. Scope
   1.2. Discrepancies
   1.3. Failure modes
   1.4. Applicable standards and specifications
2. Field panels
   2.1. Main control panel
   2.2. Junction box
3. Operator interface design conventions
   3.1. Operator actions
   3.2. Equipment and system status indications
4. Control system interfaces
   4.1. PLC to DCS communication
   4.2. Data historian interface
5. Equipment interlocks
   5.1. Emergency stop switch
6. Alarms
   6.1. System alarms
   6.2. Power loss
7. Operating mode selection
   7.1. Manual mode
   7.2. Semiautomatic mode
   7.3. Automatic mode
8. Filter equipment status
   8.1. Position sensors
   8.2. Level instruments
   8.3. Measuring instruments
   8.4. Automated valves
9. Control loops
10. Process sequences
    10.1. Main operation sequence and fast purge
    10.2. Ground
    10.3. Hold
    10.4. Setup
    10.5. Filling

10.6. Filtration
10.7. Washing
10.8. Drying
10.9. Discharge
10.10. Operator prompts
11. CIP operation or other machine/installation specifics
12. PLC to DCS information/communication

With these two documents completed, the integration portion of the project can begin. The objective will be to have the mechanical and integration portions completed on time so that acceptance testing, commissioning, and installation work can be also completed on time and within budget.

APPLICATIONS

Readers of this guide by now know that making the preliminary solid-liquid filtration selection should be followed with testing. Table 5.5 provides some general guidelines to the preliminary selection of the filter technology. The initial selection is based upon the percent (%) solids in the slurry. After this parameter, the remaining parameters may either be known if it is an existing process or will be determined during the testing.

Table 5.5 Sample Application Considerations

	Filter Press	Continuous Vacuum and Pressure	Nutsche Filter and Filter Dryer	Clarification
Solid content of the suspension (%)	5–30	10–40	10–40	<5
Maximum pressure difference	100 bar	−1 to 6 bar	6 bar	10 bar
Cake thickness (mm)	5–50	5–150	5–300	20
Average particle size	1–100 microns	1–100 microns	5–200 microns	1–50 microns
Type of operation	Batch	Continuous	Batch	Batch
Comments	Good for slow filtration and can produce dry filter cakes	Excellent cake washing and predrying	Good when reactor batch times equal to total cycle times	Disposable for low flows; candle and plate filters for large flows

CENTRIFUGAL ALTERNATIVES TO PRESSURE AND VACUUM SOLID-LIQUID FILTRATION

Although there are four driving forces for filtration—gravity, vacuum, pressure, and centrifugal—this guide focuses on the first three. However, for some applications centrifugal separation is a better approach. This section provides an overview of filtering centrifuges in terms of classification and types, when to use a centrifuge, basic process steps, and selection lab testing.

Classification (Categories) of Centrifuges

Centrifuges can be categorized by batch or continuous operation, orientation of the basket, and finally the type of cake discharge.

A continuous centrifuge accepts a continuous slurry feed and filters the solids that accumulate on the screen. A batch centrifuge accepts a fixed volume of slurry and then steps through the process cycles. For batch operations, the basket orientation can be vertical or horizontal while for continuous operation, the basket shaper can be drum or conical.

Finally, in terms of cake discharge, there are several possibilities:

Batch—vertical:	Plough/peeler
	Manual
	Lift-out bag
Batch—horizontal:	Peeler
	Inverting bag
Continuous—drum:	Pusher
Continuous—conical:	Scroll
	Vibrating cone (sliding)

When to Use a Centrifuge

As in all cases of filtration, testing is critical to centrifuge selection. A test program will check pressure filtration, vacuum filtration, and centrifugal filtration along with washing and drying to determine the optimum solution. Centrifuge testing is briefly described at the end of this section.

Generally, a centrifuge is applicable for larger and coarse solids up to 2 mm. Fine particles are normally not filtered on a centrifuge due to forming compressed cakes and blinding of the filter cloth. Filtration times are normally below 20 s, i.e., fast filtration that results directly from the particle sizes and particle shapes. In terms of particle shape, spherical, rounded, and irregular-shaped particles are more easily handled on the centrifuge. They are less compressible and do not form compact cakes under the high gravitational forces (g-forces). Needle-shaped or fragile crystals may have breaking and attrition problems at the high centrifugal speeds. Finally, abrasive solids also must be avoided on a centrifuge as they result in high wear costs of seals, filter media, and the like.

The slurry characteristics are also important to selection. A higher percent solids in the feed stream is always better and more economical; the normal range would be between 10% and up to 50% solids. If filter aid is required, then a centrifuge is not the correct piece of equipment.

Finally, the washing and drying requirements must be examined. If there are exacting wash requirements such as low conductivity or long washing times (up to several minutes), then a centrifuge is not applicable. A centrifuge will, however, most often provide a drier cake than most other filtration technologies. Obviously, there will be trade-offs between washing and drying for the process.

Basic Process Steps
There are generally seven process steps in a centrifugal batch cycle:

1. Accelerate to feed speed
2. Feeding of the suspension
3. Main filtration or spin speed
4. Washing and wash filtration
5. Dry spinning at high g-forces
6. Deceleration or slowing the basket to discharge speed
7. Cake discharge

In continuous centrifuges, the cycles are different:

1. Suspension inlet and distribution
2. First stage basket for up to 80% removal of mother liquor
3. Second stage basket
4. Product washing
5. Filtrate separation
6. Flushing

Selection Lab Testing
There are three basic laboratory tests that centrifuge manufacturers should conduct for determining the optimum design. The static settling, filtration rate, and spin settling rate tests each merit attention.

Static Settling Tests

The static settling test quickly determines if a centrifuge is possible for the process. A representative slurry sample is placed in a flask or beaker and allowed to stand for 30 min. As the solids settle, the density differences should become evident. The slightest separation would indicate that centrifugal forces can be used for separation and that further tests are required. If there is no separation, then pressure or vacuum testing is necessary.

Filtration Rate Tests

Filtration rate tests were described in Chapter 2. The analysis of the results will determine if centrifugal force or vacuum is applicable to the process filtration. If the testing shows that both are acceptable, congratulations to the entire process team for developing good crystals. Secondly, pressure filtration testing should be conducted just to ensure that all bases are covered. In the end, therefore, the other requirements of the process (washing, drying, etc.) and project (economics, layouts, operations, etc.) will determine the optimum separation equipment.

Spin Settling Rate Tests

The spin settling rate test determines the impact of the g-forces on the separation. This test is normally conducted in a bench-top test tube spinner that can produce up to 1000 Gs. The critical time is a 90-s spin for separation. If there is good separation (and this is subjective) at 90 s, then pilot testing would be recommended.

In summary, the centrifugal testing, along with pressure and vacuum testing, will narrow down the technology choices. Other process and project parameters, as previously discussed, will further dial down the process separation decision.

LIFE CYCLE CAPITAL EQUIPMENT COSTS

In the early 1980s, Edward Deming stated "organizations should end the practice of awarding business on the basis of the price tag along and, instead, minimize the total cost" (as cited in Hoffman, 2013, p. 25). This is true throughout process decisions. After all of the testing and analysis is complete, it is time to analyze the project economics. The purpose of the life cycle costs (LCC) or total install costs (TIC) analysis is to determine the most cost-efficient solution. Every operating company has a formula/plan for the process engineer to calculate return-on-investment (ROI) such as what production revenues can be credited or what costs can be saved/eliminated as well as how to calculate installation costs, etc. This section provides a summary discussion of elements to consider. Major cost components include:

- Capital
- Installation
- Operating
- Utilities
- Environmental/sustainability
- Maintenance
- Downtime
- Decommissioning
- Continuing support

The capital costs will reflect the equipment design and must be discussed at length with the technology supplier. There are basic designs as well as standard optional designs and then there are "new designs" developed for the specific process. Basic options can be reviewed in terms of the cost benefits. For instance, is an automated clean-in-place cycle required or can the cleaning be conducted manually? Remember, manual cleaning requires not only the cleaning solutions (acids, solvents, water) but how to pipe the solutions, operating costs, safety, procedural (will the procedure be followed exactly each time, e.g., day shift versus night shift). Discuss these procedures with the supplier. For example, let's say the filtration system is designed for dry cake discharge but needs cleaning periodically. How will the liquid cleaning solution be discharged, as all of the downstream piping and equipment are designed to handle dry cake but not liquids or liquids containing solids?

The "new designs" are important, as this is how machines are improved, yet they require time and money. As experience shows, every chemical engineer would like the newest and most improved filtration system, controls system, or process design. While this can be fun and exciting, it can also be a serious distraction to the project team and the supplier's team. This should be taken as a word of caution such that the project manager keeps everyone focused on the task at hand; the operating company is tasked to produce chemicals, pharmaceuticals, clean flue gas, etc., and not to design machines.

The last part of the capital cost equation is the extent of the scope of supply from the supplier. There are several alternatives: (i) filter only, (ii) filter including piping, valves, instruments, controls, and a full skid, and (iii) filter only plus engineering package from the supplier that the project team can use to engineer and fabricate the filter installation. Each alternative's pros and cons must be evaluated.

In some cases installation costs, depending upon the size and complexity of the filtration unit, may exceed the capital cost. During the building design, allowance should be included for preventive maintenance access, removal of the machine parts, overhead cranes, supports, piping/valves/instruments, etc. It is also important to determine how to get the unit into its location. Are there interferences of ceilings, doors, staircases, or platforms? Another word of caution: never dismantle a filtration system to get it into the location—it is always easier to remove the interference at the plant!

The operating, utilities, and environmental/sustainability components can be grouped together. Operating costs include material costs (such as raw materials—can the slurry be thickened to save costs?) as well as personnel labor costs. For instance, is it more efficient to design in more containment to save time and costs to have operators "suit-up" with PPEs (personal protective equipment)? Utilities, such as energy, water, or compressed air/gas, are also an important consideration. Finally, there are the environmental costs such as waste disposal and designing for sustainability (can a less hazardous solvent be used or can gas and liquid streams be recycled?).

In terms of maintenance, there is preventive (PM) and repair (normal and catastrophic). These two areas must be discussed with the technology supplier. For example, for spare parts, topics include: start-up, 2-year and capital, spare parts location and stocking, and manufactured parts versus purchased parts by the supplier. Are the service contracts available for PM training (on-site and at the factory)? Lastly, what are the risk mitigation plans for a catastrophic event? These should be all part of the overall discussions for the technology decision.

The next component is process downtime. For example, what will the supplier guarantee in terms of equipment up-time? If there is a failure, how long will it take to repair and put the unit back into service?

There are many questions that need to be considered depending upon the selected technology.

The concept of decommissioning is not one that is generally considered on a new process and new project. Every engineer believes that the current process will be in place for a while. Of course, markets change, customer requirements change, and internal and external forces change. The process engineer should be asking questions of the upper management to determine if decommissioning should be included as part of the analysis. Whatever the answer, flexibility should always be included in the evaluation of a process. The engineer might consider, for instance, the ease of adapting the filtration technology to a new process or whether the filtration design can be modified to accept different feed slurries to produce different quality filter cakes and filtrates.

Finally, there are the ongoing support questions that are very subjective in nature, for example, reliability of the parts supply; technical, mechanical, and process support; application assistance; training; and new developments. Capital purchasing similar to all purchasing, business or personal, comes down to interpersonal relationships. Visit the supplier and gauge your level of trust of supplier management. Unfortunately, there are no "consumer reports" for capital equipment; so trust the marketplace and listen to your process friends and colleagues.

Having addressed the components in filtration selection, this guide turns next to commission and operation.

CHAPTER 6

Commissioning and Operation

In planning for the process to go into operation, the process engineer must also plan for commissioning and preventative maintenance, while being ready to troubleshoot and address interesting process challenges that come to light after the fact.

COMMISSIONING PLAN

Every equipment vendor as well as operating company should have specific formats to factory acceptance testing (FAT), site acceptance testing (SAT), mechanical start-up (water batching), and then start-up. These formats and requirements will also vary according to the market place such as chemical, pharmaceutical, and oil and gas. Finally, there also needs to be a training plan for the operators.

The typical topics to be included in all of the above processes are:

1. Protocol approvals
2. Introduction
   a. Purpose
   b. Scope
   c. Responsibility
3. General
   a. Equipment description
   b. Auxiliary equipment descriptions
   c. Process functions
4. Specific reviews
   a. Nameplates
   b. Installation and lifting
   c. Fabrication, assembly, and quality control plan
   d. Material certificates and finishes
   e. Verification of P&ID
   f. Utilities
   g. Wiring

Table 6.1 Typical Training Outline

1. Filtration system
  • System principles and PFD overview
  • Design details and 3D model

2. Equipment review   • Normal operations
3. P&ID review focused on control schemes   • Hazards and items to watch

4. Filter operation
  • DCS screens and overview
  • Key interlock review
  • Start-up, shutdown, and normal operations review

5. Interfacing with other operators and other equipment

    h. Controls
    i. Seals
    j. Calibrations
    k. O&M manuals
    l. Spare parts
    m. Maintenance
    n. Cleaning
5. Operations
    a. Start-up
    b. Shutdown
    c. Normal process
    d. Mechanical
    e. Controls

One important point concerning training is to ensure that operators have both the mechanical knowledge as well as the process-operations knowledge. Table 6.1 shows a typical training outline while Table 6.2 illustrates a sample filtration quiz that might be employed after the training to be sure that all of the critical points were covered.

PREVENTATIVE MAINTENANCE

It is important for the engineer to be sure of the preventative maintenance program with the selected technology. One of the considerations for the technology section is the complexity of the process and, hence, the complexity of the filter system; a more complex process requires a more complex solution. Each equipment vendor will have a unique preventive maintenance program; it is for this reason that

Table 6.2 Sample Training Quiz

Process questions:

1. How many types of filters are there in the package?

2. How many filters are there of each type?

3. Are the filters run in parallel or in series (one feeding the other)?

4. How are the slurry and fluids fed to each type of filter?

5. What are the various "feeds" (inputs) to the package?

6. What are the various "products" (outputs) from the package?

7. How is solvent prevented from being lost with discharged solid cake?

8. What parameter stops the filtration cycle?

9. What other parameters stop a filter operation?

10. How can filtrate quality be improved?

11. How can the pressure drop across the cake be reduced?

12. How is the cake discharged from the filter?

13. Where does the cake go after discharge?

14. What are the sources of cake wash solvents?

15. Where is the filtrate sent?

16. What is the source of compressed air or gas?

Mechanical questions:

1. What parts of the filters must be changed frequently?

2. How is the filter media installed?

Controls questions:

1. How do you start and stop the filtration unit?

2. How do you start and stop pumps?

3. What are the main variables that control the efficiency of the filtration unit?

4. Who can change the control parameters?

operations/operators should also be involved in the technology decision. Tables 6.3 and 6.4 illustrate typical mechanical and process preventative maintenance programs. You can be sure, though, that there needs to be regular preventative maintenance done to reduce the regularity of troubleshooting.

TROUBLESHOOTING

There are normally three main areas that must be examined when the plant engineer receives a telephone call that the filtration system is not working. These include: (i) the filter itself for mechanical reasons, (ii) the equipment around the filter is not working, or (iii) the filter operational procedures are not correct. In examining these areas, it is

Table 6.3 Typical Mechanical Preventative Maintenance Programs	
• Check interior leaks	Daily
• Check packing glands	Weekly
• Visual checks for damage, dirt, and wear	Weekly
• Check gear box and oil levels	Monthly
• Check cake discharge	Monthly
• Check tightness settings (torque)	Monthly
• Check lubrication	Every 3 months

Table 6.4 Typical Process Preventative Maintenance Programs	
• Check filter media	Daily
• Check filter movement cylinders	Weekly
• Check proximity sensors, limit switches, etc.	Every 2 months
• Check pneumatic and electric controls	Weekly
• Check instruments and valves	Every 2 months
• Visually inspect piping	Weekly
• Check filter cake quality	During operation
• Check cleaning devices	During operation
• Check process variables	During operation

necessary to separate the symptoms from the causes. As will be seen, excessive solids in the filtrate (the symptom) can have many causes (such damaged filter cloth, PSD changes from the reactor, or post-precipitation due to temperature changes).

The first area to troubleshoot could be a failure of the equipment itself such as internal components or seals or other aspects addressed in the "Preventative Maintenance" section. The equipment itself should be the first item to be checked by the operators before calling the process engineer.

Secondly, the filtration system is part of the entire process including the upstream and downstream equipment. For example, are the reactors performing correctly in terms of agitation or temperature control in order to produce the specified crystals? Are the precoat and body feed systems in tune for mixing, feeding, flow rates, or solids loading? Are the valves and instruments operating correctly and reading the correct variables (calibrations)? Next, consider the pumps that feed the slurry and washing liquids as well as the compressors that feed the gas streams for drying and cake discharge. The pumps must

Table 6.5 Common Filtration System Symptoms
1. Filtrate is entrained with the cake discharge
2. Solids are present in the mother liquor
3. Solids are present in the wash filtrates
4. High motor torque and high amps
5. Uneven cake buildup or no cake buildup
6. Compressed cake
7. Filter media plugging or damage
8. Flow control and high or low flow rates of the slurry, washing liquids, or gases
9. No or low/slow filtration
10. Poor wash quality
11. Poor drying quality
12. Bypassing of slurry, wash liquids, or drying gas
13. No movement of gear box, coupling, cylinders, motors, VFD, etc.
14. Filter aid additions for precoat and body feed
15. Chemical addition of flocculants and coagulants
16. High differential pressures and short cycle times

produce the required pressure, flow rates, and more. The compressors must also produce a certain gas flow at a specific pressure for a certain amount of time. Finally, the engineer might explore whether the filter problem is caused by interlocks in the control system or a control communication problem.

Finally, there may be process or operational procedures resulting in filtration problems. For example, changes in PSD or amount of solids in the slurry or the cake compressibility could each cause problems. Additionally, in terms of the operation, changes in filtration pressure, timers, or speed change could have impact. Some of these changes may be caused by inadequate operator training or by a simple disregard for proper procedures, leading an operator to change a process parameter.

In summary, there are many causes of filtration symptoms that will need to be examined. Table 6.5 lists common symptoms; the causes are more difficult to uncover and may require additional lab testing to resolve.

INTERESTING PROCESS CHALLENGES AFTER THE FACT

Throughout, this guide has noted how upstream and downstream equipment and processes impact the solid-liquid filtration system.

This section will relate real-world examples of what appear to be small changes that are, in fact, large changes impacting the filtration system. The examples include a nutsche filter dryer, continuous rotary pressure filter, continuous vacuum filtration, candle filtration for clarification, pressure plate filtration with precoat and body feed, and an optimization to replace candle filters with continuous-indexing vacuum belt filters. The take-away from these examples is that engineers, during process filtration troubleshooting, should always look upstream and downstream before "jumping to conclusions" about the filtration system.

Nutsche Filter

In one pharmaceutical case for a nutsche filter, a small change in the reaction chemistry, adding a sodium ion to a polymer chain, resulted in a major filtration problem. The product, during the drying phase, formed "balls" and entered into a thixotropic phase. The process engineers could not explain the phenomena until a chemist explained this seemingly inconsequential change. The resulting fix required changes to the motor and torque requirements.

Continuous Rotary Pressure

In a specialty chemical application for a continuous rotary pressure filter, testing showed a dual distribution in the product's PSD. The sampling was conducted on the batch reactor. The engineering team, as part of the project to install a continuous process, converted the batch reactor to a continuous reactor and assumed the same PSD. Unfortunately, the continuous reactor produced a completely different PSD, which then required changes to the filter media.

In another specialty chemical application, the testing showed that continuous rotary pressure filter would achieve the requirement for filtration, cake washing, and drying. After installation and start-up, the filtration flux rates were completely different, much lower, than the testing with the result that the continuous process steps could not be realized. The engineers examined all of the process parameters without success. After reviewing all of the data including reaction chemistry, it was determined (after several months) that the zeta potential (ionic charge) of the slurry had changed due to the process flow and that agglomeration was occurring in the filtration system within the filter itself. A small change to the pH finally corrected the process problems while not impacting the reaction chemistry.

Clarification Application with Candle Filters

In a clarification application with candle filters, the testing resulted in a duplex candle filter skid system with each candle filter having 25 m^2 of filter area. The system was installed and started up and successfully ran for over 1 year. Inexplicably, the performance changed drastically and the filter media began plugging very quickly during the cycle. The process engineers explained that there had been no change in the process and began to investigate the upstream conditions (Figure 6.1).

Filter A			
	Start	End	Total
Liquid flow (m³/h)			
Suspended solids (mg/l)			
Temperature (°C)			
pH			
Pressure (bar g)			
Batch time (min)			

Filter B			
	Start	End	Total
Liquid flow (m³/h)			
Suspended solids (mg/l)			
Temperature (°C)			
pH			
Pressure (bar g)			
Batch time (min)			

Clarifier			
Feed to clarifier: pH			
Feed to clarifier: temperature			
Coagulant (type)			
Feed coagulant (g/l)			
Flocculant (type)			
Feed flocculant (g/l)			
Feed slurry to clarifier (m³/h)			
Suspended solids (mg/l)			

Oxidation unit			
Running	Yes	No	

Figure 6.1 Clarification application with candle filters.

After months of work, the process engineers narrowed down the changes to the chemical supplier of the flocculants and coagulants. The vendor and the client began lab testing again using the initial slurry along with the chemical supplier and their flocculants and coagulants. The resulting testing showed that the chemical change caused the larger particles to settle out quickly and therefore only the smaller particles reached the filtration system. The client and the chemical company reduced the amount of chemicals used (better for the client and not so good for the chemical supplier), which produced a smooth PSD and better filtration rates. Once again, we see the systems approach to process filtration and "not jumping to conclusions" as the guiding principle.

Continuous Vacuum Belt Filter

In this dual stage bioenergy process, two continuous vacuum belt filters are installed. For the first part, hemicellulose, the filtration times are very fast and a 40-mm cake is achieved in 15 s. The subsequent washing steps each require about 4 s. The filter media is polypropylene, 20 microns, and there is no blinding and visually clear filtrate.

In the second stage, there is a fast degradation of the filter media. Laboratory tests were undertaken to determine the process question. During the initial testing on the 20-micron media, there were visible solids blinding. It was hypothesized that the stripping processes reduced the particle size to a point that this media is not a good choice.

To understand the issues, the media was treated with a wash of NaOH overnight. It seemed that there was precipitation taking place inside the filter cloths when the pH dropped. The cloth was able to be completely rejuvenated to achieve repeatable filtration times with long soaks (15 min) and gentle rinsing. For the production unit, this indicated that a caustic cloth rinse after solids were discharged was required to renew the cloth. The new design included an aggressive, high-pressure spray to renew the cloth on each pass. The final design had the caustic wash, containing solids, directed back to the slurry feed tank. The benefit was that the net usage of caustic for the process would not be impacted as that NaOH would go toward the amount fed to the reslurry tank.

Secondly, new filter media was tried and a tighter weave was selected to allow for cake building on top of, and not within, the media. The times for filtration on rejuvenated cloth were in the initial range of 10−15 s for a 10-mm cake with the cake washing in the 4-s range.

Finally, there was some concern about the impact of a water wash and acid neutralization on filtration time. These both reduce pH and can cause precipitate plugging. Additional testing showed that with a slight caustic wash, the cake wash would have no issues. Therefore, in this case, thorough laboratory testing resulted in a new filter cloth, different cloth washing, pH control, and thinner cakes.

Pressure Plate Filtration Systems

In this specialty chemical process, there are two pressure plate filtration systems installed—one with centrifugal cake discharge and one with vibratory cake discharge. The process has several stages:

1. Precoat (1 h): The filter is precoated using Hyflo Supercel.
2. Filtration with body feed (2−5 h): The body feed is the same as the precoat (Hyflo Supercel). However, when a reaction performs badly, there is an increase in the amount of solids (fines) and the filtration has gone as long as 11−15 h and produced cake >60 mm. In some instances, this completely packs the gap between the filters.
3. Organic solvent washing (2−3 h): This is a continuous wash rather than a displacement wash. Due to the nature of the cake, this is a very slow process and not as effective as it should be with standard displacement washes.
4. Nitrogen drying (5 h): This is "shock" drying to about 50% moisture.
5. Steam stripping (time is variable): This step removes the residual solvent. As the solvent is removed, the product precipitates and "glasses" the cake, which makes discharge more difficult.
6. Nitrogen drying: This step is repeated to remove the residual water from saturated steam.
7. Discharge: The cake discharge is centrifugal or vibratory.
8. Water cleaning: Water is added and allowed to soak, drain, and dry, and then discharged to remove the residual cake from the media.

In this case, the objective for the laboratory testing to improve the process was to increase capacity and process reliability, improve filtrate clarity, and achieve a drier cake for increased product yield.

The testing improved the process with a change to the filter aid for the precoat and body feed such that there was increased precoat and lower body feed. Secondly, the body feed was made a different grade than the precoat to allow for improved cake formation. The filter media was changed to allow for a clean filtrate. Finally, the improved filter aid and filter media selections improved the reliability, reduced the cycle times, and improved product yield. In summary, therefore, an improved cake structure solved the process problems.

Replacing Candle Filters with Continuous-Indexing Vacuum Belt Filters

In this specialty acid application, the process engineer initially selected candle filters for the pilot plant, which was, as described below, an incorrect choice. The process included filtration, water washing, neutralization, and drying. Unfortunately, there was an immediate "jumping to conclusions" without a full evaluation. The decision for candle filters was based upon a vendor recommendation with minimal lab test data as well as economic and delivery objectives.

In the pilot plant, there was difficulty in drying and cake discharge as well as cake cracking. As seen earlier, some cakes cannot be vertical. In this case, the cake structure was fragile and did not hold its structure well in a vertical orientation. The learning from the pilot plant included that the vertical cake structure was hard to maintain, drying was limited to 80% solids, and a wash ratio of 10:1 was needed to meet the quality specification. As a result of the pilot plant work, it became clear the candle filter was not the optimal technology.

For the commercial plant, a full lab evaluation was conducted to include filter media selection, filtrate clarity, washing, neutralization, and drying by vacuum including blowing and pressing. The decision for the commercial plant was for a continuous-indexing vacuum belt filter with the benefits of horizontal cake structure, multiple washing steps, multiple neutralization steps with full displacement washing, clear filtrates with no detectable solids, and improved drying to >90% solids. Finally, an added benefit of pressing the cake was a 40% reduction in cake thickness to allow for easier disposal and to meet the EPA Paint Liquids Test (see Appendix).

The life of a process engineer is seldom an easy one. There are so many factors to consider. It's the challenge that keeps it interesting.

CHAPTER 7

Conclusion

This practical guide's one aim has been to help the process engineer better design a filtration process. To best illustrate how an engineer might draw together all of the information covered over these pages, let's consider one last process problem.

In this case, the process problem is to design a filtration system to extract the liquid ion product from a zeolite catalyst. The production requirement is 10 tons/hour on a continuous basis. The first question that arises is what is the meaning of 10 tons/hour? Is this on a dry solids basis or a slurry basis? The feed slurry consists of a water/solid mixture that is 50%/50% on a weight basis. The particles have a dual PSD with the $d_{50} = 60$ microns and a PSD tail of <7 microns.

Based on the discussion in Chapter 3, the engineer has hypothesized that for continuous filtration there are three choices: rotary pressure filter, rotary vacuum filter, or horizontal vacuum belt filter. Following this, the engineer heeds this guide's recurring advice to test, test, test! The next step is to go to the lab. The chemist uses the trusty Buchner funnel and makes up several "representative slurries" to gather the initial data. This good start nets the following results.

The first test used feed at 200 g of solids + 200 ml of water. The Buchner funnel has a diameter of 125 mm, and 159 ml of water was filtered in 90 s with a final cake thickness of 19 mm. The second test used feed (fines) at 200 g solids + 400 ml of water. In this test, 307 ml of water was filtered in 255 s with a final cake thickness of 17 mm.

These initial screening tests show that vacuum filtration is possible. However, as discussed in Chapter 2, much more data is required. Table 7.1 shows the necessary data for a complete analysis, and Table 7.2 shows a typical setup.

The testing then needs to continue to examine different wash ratios, countercurrent washing, drying and drying techniques (such as

Table 7.1 Typical Test Data Required

Product:			Test Number:
Date:			
		Units	Run #
	Filter media		
	Suspension		
	Sealing gas flow		
	Sealing vacuum		
Filling	Volume of slurry		
	Weight of slurry		
	Density of slurry		
	% Solids in feed		
Filtration	Vacuum		
	Gas flow		
	Temperature		
	Volume of filtrate		
	Time for filtration		
	% Solids in filtrate		
Wash 1	Wash material		
	Vacuum		
	Gas flow		
	Temperature		
	Volume of filtrate		
	Time for filtration		
	% Solids in filtrate		
Wash 2	Wash material		
	Pressure		
	Temperature		
	Volume of filtrate		
	Time for filtration		
	% Solids in filtrate		
Wash 3	Wash material		
	Pressure		
	Temperature		
	Volume of filtrate		
	Time for filtration		
	% Solids in filtrate		

(Continued)

Table 7.1 (Continued)			
Product:			**Test Number:**
Date:			
		Units	**Run #**
Drying	Vacuum		
	Gas flow		
	Temperature		
	Time for drying		
Cake	Wet cake weight		
	Thickness		
	% Residual moisture		
	Discharge ok?		
	Cake rests on filter cloth?		
	Tare weight, g		
	Total dry weight, g		
	Dry solids, g		

Table 7.2 Test Setup

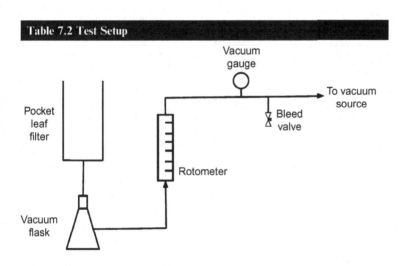

vacuum only, vacuum and blowing gas, vacuum and blowing gas, and mechanical pressing), as well as temperature variations to maximize the product yield.

Unsurprisingly, for anyone who has read this guide from the beginning, testing is always essential. A typical test plan is included in Table 7.3 for pressure filtration using filter aid. Filtration process

Table 7.3 Typical Testing Plan for Pressure Filtration Using Filter Aid	
A. Objectives	
1)	Improved reliability	
2)	Improved clarity	
3)	Improved capacity	
B. Key Metrics/Variables	
1)	Filter media:	Select micron rating appropriate for precoat formation
2)	Precoat:	Select precoat to generate clarity
		Reduce precoat build time
3)	Filtration pressure:	Improve flux to reduce cycle time
4)	Body feed material:	Improve flux to reduce cycle time
		Improve washing to increase yield and reduce cycle time
5)	Body feed amount:	Improve flux to reduce cycle time
		Improve washing to increase yield and reduce cycle time
6)	Cake thickness:	Improve flux to reduce cycle time
		Improve washing to increase yield and reduce cycle time
C. Base Case	
1)	Media:	80 micron SS woven media
2)	Precoat and body feed:	Hyflo Supercel
3)	Filtration:	Slurry to be provided
		45 psig
		Approx. 40 mm cake
4)	Solvent washing:	Toluene
		Use different wash liquid to solids ratio
		Analyze samples for product concentration
5)	Drying:	Confirm airflow and drying time
6)	Drying:	Confirm airflow and drying time
D. Testing Variables	
1)	Filter media:	80, 40, and 25 microns
2)	Precoat:	Hyflo Supercel and Celite 577
3)	Filtration pressure:	45 and 90 psig
4)	Body feed material:	Hyflo Supercel and Celite 545 at different feed amounts
5)	Cake thickness:	40 and 20 mm

(Continued)

Table 7.3 (Continued)		
E. Plan Execution		
1)	Initially start with the base case settings but run thinner cake thickness to build the filtration curve	
	a. Test at 10 and 20 mm	
	b. Extrapolate to 40 mm and compare to process results	
2)	Run Celite 545 at 8%	
	a. Set: 40 micron media, Supercel precoat, 45 psi pressure	
	b. Run at 20 mm cake thickness to start	
3)	Run Celite 545 at 32%	
	a. Set: 40 micron media, Supercel precoat, 45 psi pressure	
	b. Run at 20 mm cake thickness (will be slightly higher)	
4)	Run Celite 545 at 16%	
	a. Set: 40 micron media, Supercel precoat, 45 psi pressure	
	b. Run at 20 mm cake thickness (will be slightly higher)	
5)	Run Celite 545 at a higher amount if deemed necessary	
6)	Determine best case and try higher pressure	
7)	Using best case, verify clarity and switch to tighter precoat if needed	
8)	Notes for washing: In each test case, examine the washing efficiency by looking at both time and amount. This can be easily done by taking 3 equal wash amounts as indicated previously, timing each, collecting each, and testing for product in the filtrate. The final cake can also be tested for product.	
9)	Notes for drying:	
	a. After determining the optimum filtration and washing and filter aid usage, then begin the drying testing	
	b. Test at blowing pressures of 30, 60, and 90 psig	
	c. Test blowing times of 60, 90, and 120 s	
	d. After the optimum is determined, check the impact of the gas temperature	

development requires a good test plan written specifically for the application. A test plan will lead you into creativity necessary for process development. Recalling the wisdom of Sherlock Holmes and Dr. Watson in Chapter 1, we know there is no benefit to jumping to conclusions. With conscientious attention to the many factors in filtration, we all must learn to tell the crucial from the incidental.

Paint Filter Liquids Test

In Chapter 3, there was a discussion of candle filters and plate filters. In applications where the filtrate is the product and the solids are for disposal, a standard specification is "No free liquids" using the Paint Filter Liquids Test (Method 9095B), which is in compliance with 40 CFR 264.314 and 265.314. A summary of the test method is appended for the process engineer's reference.

1.0 Scope and Application
    1.1. This method is used to determine the presence of free liquids in a representative sample of waste.

2.0 Summary of Method
    2.1. A predetermined amount of material is placed in a paint filter. If any portion of the material passes through and drops from the filter within the 5-min test period, the material is deemed to contain free liquids.

3.0 Interferences
    3.1. Filter media is observed to separate from the filter cone on exposure to alkaline material. This is not a problem if the sample is not disturbed.
    3.2. Temperature can affect the test results if the test is performed below the freezing point of any liquids in the samples. Tests must be performed above the freezing points and can exceed room temperature of 25°C.

4.0 Apparatus and Materials
    4.1. Conical paint filter: Mesh number $60 \pm 5\%$
    4.2. Glass funnel
    4.3. Ring stand
    4.4. Graduated cylinder or beaker with 100 ml capacity

5.0 Reagents: None

6.0 Sample Collection, Preservation, and Handling
A 100 ml or 100 g representative sample is required for the test. If it is not possible to obtain a sample of this size, the analyst may use larger sample sizes in multiples of 100 ml or 100 g. However, when larger samples are used, the analyst shall divide the samples into 100 ml or 100 g lots and test each lot separately. If any lot test contains free liquids, then the entire sample is considered to have free liquids.

7.0 Procedure

7.1. Assemble per Figure A.1.

7.2. Place the sample in the filter. Settling the sample in the paint filter may be facilitated by lightly tapping the side of the filter as it is being filled.

7.3. Allow the sample to drain for 5 min into the graduated cylinder.

7.4. If any of the test material collects in the graduated cylinder in the 5-min period, then the material is deemed to contain free liquids for the purposes of 40 CFR 264.314 and 265.314, per Figure A.2.

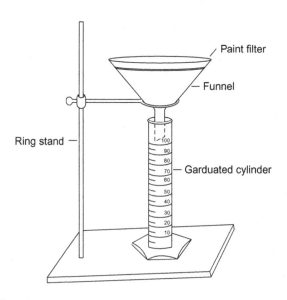

Figure A.1 Paint filter test apparatus.

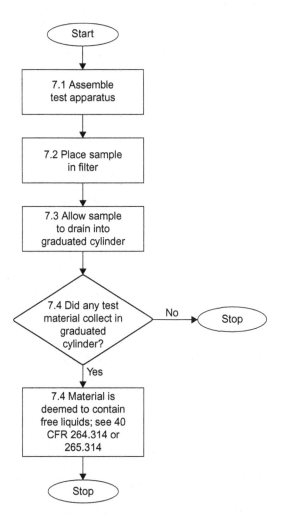

Figure A.2 Paint filter liquids test.

A

Absolute A degree of filtration that guarantees 100% removal of suspended solids over a specified size found in the filtrate.

Absorption The taking in, incorporation, or reception of gases, liquids, light, or heat. Penetration of one substance into the inner structure of another, using filling the void of the matrix. The process of movement of a drug from the site of application into extracellular compartment of the body.

Activated carbon Charcoal activated by heating to 1472–1652°F a material of high adsorptive gases, vapors, organics, etc. Has a large internal surface area. Removes dissolved color, odor, and taste from liquids or gases. Commonly used in the pharmaceutical industry to remove organic contaminants.

Adsorption The adhesion of a substance to the surface of a solid or liquid. Adsorption is often used to extract pollutants by causing them to be attached to such adsorbents as activated carbon or silica gel.

Agglomeration, particle Multiple particles joining or clustering together by surface tension to form large particles, usually held by moisture, static charge, or particle architecture.

B

Backpressure A backward surge of pressure from downstream to upstream of the filter. Can be the result of closing a valve or air entrapped in a liquid system.

Backwash Reversal of a fluid flow through the filtration media to remove solids from the filter. Intended to clean or regenerate a filter.

Beta ratio Measurement of filter retention efficiency. Ratio of particles exposed to a filter, as a feed stream to the particles downstream (filtrate).

Blinding Blockage by dust, fume, or liquid not being discharged by the cleaning mechanism; results in a reduced gas or liquid flow of increased pressure drop across the filter media.

Breakthrough Used to describe the passing of solids through the cake buildup of a filter medium. Also called breakpoint.

Bridging Material or particulate blockage across an opening, often of a pore or filter medium.

Brownian motion The continuous zigzag motion of suspended minuscule particles. The motion is caused by impact of the molecules in the fluid upon the particles.

Bubble point pressure A test to determine the maximum pore size openings of a filter. The differential gas pressure in which a wetting liquid (e.g., water) is pushed out of the largest pores and a steady stream of gas bubbles is emitted from a wetted filter under specific test conditions. A filter integrity test with specified, validated pressure values for specific pore size and type filters.

C

Cake (filter) Solids deposited on the filter media. In many cases, the cake may serve as its own filter medium.

Cake release Ability of a medium to allow clean separation of the cake from the medium.

Calendaring A manufacturing process where woven and/or nonwoven fabrics are pressed between heavy rollers compressing the fibers. The process reduces the filter medium void volume, pore size rating, flow rate, and dirt-hold capacity of the medium.

Candle filter A reusable filter consisting of a tube made from ceramics or metal. Flow is from the outside-in with particulate accumulating on the outside of the candle. The candle can be cleaned by various means, including back-pulsing, heat, or chemicals.

Capacity Volume of product a housing will accommodate expressed in gallons or similar units. Also, amount that will filter at a given efficiency and flow rate typically expressed in gallons per minute or similar units.

Clarification Clearing a liquid by filtration, by the addition of agents to precipitate solids, or by other means.

Clarity Amount of contaminate left in a filtered liquid.

Classification Condition in which larger particles settle out below the finer ones. Also referred to as stratification. May also be referred to as the action to sort out particles by various groups or to other established criteria.

Cleanability The ability of a filter element to withstand repeated cleanings while maintaining adequate dirt capacity.

Clean pressure drop Differential pressure (drop) across measured in pounds per square inch at rated flow on new elements with clean product.

Coagulation In water and wastewater treatment, the destabilization and initial aggregation of colloidal and finely divided suspended matter by the addition of a floc-forming chemical or by biological processes.

Cold sterilization Removal of all bacteria by filtration through a sterilizing grade 0.2 μm absolute filter.

Collapse pressure The outside-in differential pressure that causes the structure of a filter medium failure of a filter element.

Collection efficiency Percentage of contaminate collected.

Colloid Very small, insoluble, nondiffusible solid or liquid gelatinous particles that remain in suspension in a surrounding liquid. Solids usually on the order of 0.2 μm or less.

Compressibility Degree of physical change in filter cake particles when subjected to normal pressures.

Contaminate Unwanted foreign matter in a fluid accumulated from various sources such as systems dirt, residue from moving parts, or atmospheric solids.

Continuous phase Basic product flowing through a filter or filter separator, which continues on through the system after being subjected to solids and/or other liquid separation.

D

De Diatomaceous earth. A filter aid from diatomites.

Dead end filtration Feed stream flows in one direction only—perpendicular to and through the filter medium—to emerge as product or filtrate.

Depth filtration A process that entraps contaminants both within the matrix and on the surface of the filter media.

Dewatering A physical process that removes sufficient water from sludge so that its physical form is changed from essentially that of a fluid to that of a slurry or damp solid.

Diatomaceous earth filtration (D.E.) A filtration method that uses a medium consisting of microscopic shells of single-celled plants known as diatoms.

Diatoms Group of phytoplankton with silica present in the cell walls.

Differential pressure—Delta (Δ) P The change in pressure or the pressure drop across a component or device located within the air stream; the difference between static pressure measured at the inlet and outlet of a component device.

Diffusion In liquid cake washing, removing the original liquor around the individual particles by mixing with the wash liquor. In air, the particle at a size within 1 or 2 orders of magnitude of the gas flow molecules moves in Brownian motion and collides with a fiber or other filter media material during its random path of travel.

Diffusional interception In gas filtration, at low gas flow velocities, tiny particles are subject to Brownian motion, enabling them to move out of the gas streamlines and become intercepted by the filter.

Direct interception Gas filtration: particles larger than the pores are removed by direct contact with the filter surface. Some particles smaller than pores can be removed as well depending on the proportion to their size hitting the surface.

Dirt (holding) capacity Amount of dirt or debris retained by a filter in grams per unit area of the filter medium.

Discontinuous phase Separated phase or product from the continuous phase. Example: Water may be in the discontinuous phase when separated from hydrocarbon, air, or gas.

Dissolved solids Any solid material that will dissolve in a liquid such as sugar in water.

Durometer (shore) Measure of hardness. Must be defined as being either A or D scale.

E

Effective filtration area The portion of filter that fluid flows through during the filtration process.

Efficiency Degree to which a filter device will perform in removing solids and/or liquids.

Effluent Fluid that has passed through a filter (filtrate or product stream). Outflow from other treatment such as wastewater treatment plants.

Entrainment To carry suspended particles within a fluid stream.

Extractables Chemicals leached from a filter during a filtration process. Usually test for extractables by soaking in water under controlled conditions. May be removed by pre-flushing with suitable liquid.

F

Feed Materials to be filtered. Also referred to as concentrate, influent, intake, liquor, mud, prefilter, pulp, slime, or sludge.

Fiber Any particle with length ≥ 0.5 micron and at least five times greater than its diameter, leaving substantially parallel sides.

Fiber migration Downstream migration of fibers from a filter medium.

Filter aid Small size particle substance of low specific gravity that remains in suspension when mixed with a liquid to be filtered. Increases filtration efficiency of a feed when deposited on a septum by forming a porous cake.

Filter cake The accumulation of particulate or solids on a surface. Can also mean a precoat for filtering.

Filter efficiency A measurement of how well a filter retains particles. The percentage retention of particles of a specific size by a filter.

Filter life Measure of a filter's useful service life based on the amount of standard contaminate required to cause differential pressure to increase to an unacceptable level, typically 2−4 times initial differential pressure or 50−80% drop in initial flow or the downstream measure of unacceptable particulate.

Filter media migration Problem caused by a filter medium constructed of a noncontinuous or fibrous matrix. Portions of the filter change structure and cause fibers to migrate downstream.

Filter medium Permeable material that removes particles from fluid being filtered.

Filtrate The end product of the filtration process. The liquid exiting the filtrate outlet.

Filtration Removal of particles from a fluid by passing the fluid through a permeable material.

Filtration rate The volume of liquid that passes through a given area of filter in a specific time.

Fines Portion of a powder-like material composed of particles smaller than the size specified.

Flocculation Growing together of minutely sized particles to form larger ones, called flocs.

Flow decay Decrease in flow rate caused by filter plugging or clogging.

Flow decay test Determines flow rate and throughput of a filter type or combination of filters on a specific liquid, usually by using small area filters, to determine the sizing of a filter system.

Flow rate The speed at which a liquid flows and is measured in gallons or liters per minute. Flow rate of a liquid can be affected by the liquid's viscosity, differential pressure, temperature, and type of filter used. Measuring air diffusion.

Flow resistance Resistance offered by a filter medium to fluid flow.

Fluid Includes liquids, air, or gas as a general term.

Flux Measure of the amount of fluid that flows through a filter, a variable of time, the degree of contamination, differential pressure, total porosity, viscosity, and filter area.

Forward flow test An integrity test measuring air diffusion at a low pressure (approximately 5 psi). Similar to a pressure hold test.

G

Gelatinous Used to describe suspended solids that are slimy and deformable, causing rapid filter plugging.

Gradient density A stratified cross-section. Used to describe a filter medium where larger pores are at the upstream side of the medium with finer pores downstream. The configuration increases dirt-holding capacity and improved filter life. The medium may be inverted when a surface filter effect is desired, resulting in lower differential pressure across the medium than if the medium has a single density throughout.

H

Housing A metal or plastic tank or tube with an inlet and outlet containing filter(s), allowing for the flow of a fluid and contaminate through the filter while containing the process.

Hydrometer An instrument used to measure the density of a liquid.

Hydrophilic Water accepting or wetting.

Hydrophobic A membrane or other material that repels and cannot be wetted by aqueous and other high surface tension fluids. When pre-wetted with low surface tension fluid, such as alcohol, the filter will then wet with water.

I

Impermeable Material that does not permit fluids to pass through.

Impingement Process of removing liquid or solid contaminate from a stream of compressed air or gas by causing the flow to impinge on a baffle plate at high velocity.

Inert Chemical inactivity. Unable to move. Totally un-reactive.

Inertial impaction The particle, due to its inertia and usually in stream-line flow, deviates out off the air/gas stream striking a fiber or other material of a filter medium. This is effective primarily for particles about 0.3 μm and larger.

Influent Fluid entering the filter.

Inlet pressure Pressure entering the inlet side of the filter. Also called upstream pressure or line pressure.

Inline filter A filter assembly in which the inlet, outlet, and filter element are in line.

Inorganic turbidity The portion of turbidity (light scattering in a water sample) resulting from inorganic (i.e., not containing the element carbon) particles.

In situ Sterilization or integrity testing of a filter in the system rather than as an ancillary operation such as in autoclave or bubble point stand.

Integrity test Used to predict the functional performance of a filter. The valid use of this test requires that it be correlated to standardized bacterial or particle retention test. Examples: Bubble Point Test, Diffusion Test, Forward Flow Test, Pressure Hold Test.

Interfacial tension Measure of miscibility or solubility of the continuous and discontinuous phases. Increases as miscibility or solubility decreases.

Interstices Spaces or openings in a filtration medium. Also referred to as pores or voids.

Interstitial Pertaining to the openings in a filtration medium.

L

Line pressure Inlet pressure, upstream pressure. The pressure in the supply line.

Liquor Material to be filtered. Also referred to as concentrate, feed influent, intake mud, prefilter, slime, or sludge.

Low interfacial tension Where the interfacial tension of one liquid over the other liquid would be <25 dynes/cm at 70°F.

M

Mass distribution Relative frequency distribution of mass within a particle size distribution. Sometimes presented as cumulative percentage undersize.

Mean efficiency rating The measurement of the average efficiency of a filter medium using the multi-pass test where the average filtration (BETA) ratio equals 2.0.

Mean flow pore measurement Calculated as the diameter of the pore of a membrane partially voided of liquid such that air flow of the partially wetted membrane is equal to 1/2 the dry air flow (theoretical diameter of the mean pore).

Media Material through which fluid passes in the process of filtration and retains particles. Also, nutrients containing solutions in which cells or microorganisms are grown.

Media migration Migration of materials making up the filter medium may cause contamination of the filtrate.

Medium Principle component of a filter element. Material of controlled or uncontrolled pore size or mass through which a fluid stream is passed to remove foreign particles held in suspension or to repel droplets in the case of coalesced water.

Mesh A term referring to a woven filtration medium, typically wire cloth or monofilament woven fabric.

Mesh count Number of openings or fractions of openings in a lineal inch of wire cloth or monofilament woven fabric.

Microfiltration (MF) Used for clarification, sterilization, to detect or analyze bacteria and other organisms and particulate matter. Separation of particles ranging from 0.1 to 10 μm.

Micrometer (μm) Micron, 1/1,000,000 of a meter. 40–60 μm is approximately the diameter of a human hair.

Micron (μm) The common unit of measurement in the filtration industry is the micron or micrometer. One micron equals 40 millionths of an inch (0.00004) or, expressed differently, 25.4 microns equals 0.001 in.

Micron rating The smallest size of particles a filter can remove.

Migration Contaminate released downstream of a filter.

Minimum bubble point pressure A diffusional flow pressure just before the onset of bulk flow.

Minimum critical bubble point pressure A filter specification derived from diffusional flow, bubble point curves for many filters.

Monofilament Single, large continuous filament of a synthetic yarn. Similar to fishing line in cross-section.

Monofilament woven fabric Woven fabric from monofilament yarns used as a screen or surface filter. Often used in sifting, belting, medical filters, etc. Most common yarns are from polyester, polypropylene, and nylon.

Multifilament A number of unbroken continuous fiber strands that run parallel to form a yarn. Typically used to manufacture a woven or knit fabric.

N

Needle felt A nonwoven fabric where stable fibers are entangled together through a manufacturing process using barbed needles, providing for a heavy-weight filter fabric, which can filter air-borne particles for use in bag houses and suspended particles in liquids from lighter weight needle felt fabrics for use in liquid bag filtration.

Nominal An arbitrary term used to describe the degree of filtration and generally not comparable or interchangeable between products or manufacturers. A user should always ask for a copy of the test procedure used and results from the manufacturer's lab notebook to understand each rating.

Nominal filtration rating An arbitrary micrometer value indicated by the filter manufacturer. The same ratings from two manufacturers are often different and rarely can be compared.

Nonwoven A filter fabric that is formed of natural or synthetic fibers randomly oriented in filtration media. Typically, held together with a binder or fibers are entangled.

O

Open area Pore area of a filter medium, often expressed as a percentage of the total area.

Outlet pressure Downstream pressure. Pressure exiting the outlet side of the filter.

P

Parallel filtration Branching a filtration setup. Two assemblies of the same pore size are in parallel to increase flow rate or simplify filter changes.

Particle Unit of material structure; a mass having observable length, width, thickness, size, and shape.

Particle count Practice of counting particles of solid matter in groups based on relative size contained in a certain area.

Particle size distribution The size range and quantity of particles measurable in dry or liquid sample. Used to determine the appropriate filter media for a specific process.

Particulate Any solid or liquid material in the atmosphere.

Particulate unloading The process whereby a filter, particularly a depth filter, can become blocked with particulate matter and subsequently release part of this matter downstream.

Perlite Material similar to volcanic glass with a concentrated shell structure. Used as a filter aid.

Permeability A measure of fabric porosity or openness, expressed in cubic feet of air per minute per square foot of fabric at a 0.5" water column pressure differential in air or by specified conditions for liquid.

Permeable Material that has openings through which a liquid or gas will pass in filtering.

Permeate The fluid that passes through a membrane. A term usually used with ultrafiltration or R/O.

Plugging Filtered-out particles filling the openings (pores) in a medium to the extent of shutting down the flow of a fluid. Also referred to as blinding or blocking.

Point-of-use filters Filters located immediately prior to where a clean effluent is required in a process.

Pore Opening in a medium. Also referred to as interstices. Size and shape of the openings are controlled by the manufacturer of the filter medium.

Pore size Diameter of pore in a filter medium.

Pore size distribution Exclusive to permeable medium: describes the number of pores in various groups of sizes in a way similar to that discussed under particle size distribution.

Pore size-absolute rating The rated pore size of a filter. Particles equal or larger than the rated pore size are retained with 100% efficiency.

Pore size-nominal rating The pore size at which a particle of defined size will be retained with efficiency below 100% (typically 90–98%). Rating methods vary widely between manufacturers.

Porosity The percent of open areas per unit volume of a medium whether it be a filter cake or roll stock, such as a paper, membrane, woven textile, or nonwoven fabric.

Porous metal Finely ground chards of sintered metal that serve as a filter medium. Often used in high pressure and/or temperature applications.

Precoat A deposit of material (usually inert), such as a filter aid on a septum prior to beginning filtration.

Prefilter Filter for removing gross-size contaminate before the product stream enters a finer-rated filter.

Pressure differential Difference in pressure between two points.

Pressure drop (ΔP) Difference in pressure between two points.

Pressure drop, clean Differential pressure (drop) across a housing measured in psi at rated flow on new elements with clean product.

Pretreatment Changing the properties of a liquid-solid mixture by physical or chemical means to improve its filterability.

Pulsing backflow Intermittent, on–off blowing with or without cake discharge.

R

Rated flow Normal operating flow rate at which a product is passed through a housing. Flow rate that a housing and medium are designed to accommodate.

Reentrainment Process of rendering particles airborne again after they have been once deposited from an air stream.

Residual dirt capacity The dirt capacity remaining in a service-loaded filter element after use. But, before cleaning, measured under the same conditions as the dirt capacity of a new filter element.

Residue Solids deposited upon the filter medium during filtration in sufficient thickness to be removed in sizeable pieces. Sometimes referred to as a cake or discharge solids.

Retention Ability of a filter medium to retain particles of a given size.

Reusable filters Filters that are washed or cleaned of contaminate, either in situ or off line, for additional uses.

S

Sedimentation Action of settling of suspended solids.

Self-cleaning Filtering device designed to clean itself by the use of a blowdown or backwash action.

Separation Action of separating solids or liquids from themselves (e.g., by size, viscosity, density, charge, etc.) or liquids or gases from fluids.

Serial filtration Filtration through two or more filters of decreasing pore size, one after the other, to increase throughput, filtration efficiency, or to protect the final filter.

Service life Length of time an element operates before reaching an unacceptable benchmark, e.g., maximum allowable pressure drop.

Shell Outer wall of a housing. Also referred to as the body of a housing.

Sieve A screen filter with straight-though capillary pores and identical dimension.

Silting index Measurement of the tendency of a fluid to cause silting in close tolerance devices as a result of fine particles and gelatinous materials being suspended in the fluid. Measured by a silting index apparatus.

Single pass This test system is designed to be representative of a typical filter circuit. Fresh contaminates are introduced in a slurry form into the test reservoir, mixed with the fluid, and pumped through the test filter. The test is run in such a manner to produce one pass of all fluid and contaminate.

Sintering A process of heating materials (e.g., metal or ceramic) to elevated temperature causing mating surfaces to fuse as one.

Size distribution Proportion of particles of each size (by mass, number, or volume) in a powder or suspension.

Slurry Suspension to be filtered or dewatered.

Solids Mass or matter contained in a stream. Considered an undesirable discontinuous phase and should be removed.

Solute Liquid that has passed through a filter. Also referred to as discharge liquor, effluent, filtrate, mother liquor, or strong liquor.

Solution Single-phase combination of liquid and nonliquid substances of two or more liquids.

Stream Term sometimes used and synonymous with the words product, liquid, air, gas, fluid, etc. in speaking of any matter processed by filtration or separation equipment.

Substrate Substance or basic material as a filter media or to which a deposit is added.

Surface filter Filter medium that retains particles wholly on the surface and not in the depth of the cross-section of a filter medium, e.g., plain weave wire cloth and monofilament woven fabrics or membrane.

Surface filtration A process that traps contaminants larger than the pore size on the top surface of the filter, usually a membrane, wire cloth, or monofilament fabric. Contaminants smaller than the specified pore size may pass through the medium or may be captured within the medium by some other mechanism, such as surface affinity, turboelectric potential, or other means, which prevents particle penetration.

Surface tension Tendency of the surface of a liquid to contract to the smallest area possible under existing circumstances.

Surfactant A soluble compound that reduces the surface tension of a liquid, or reduces interfacial tension between two liquids or between a liquid and a solid.

Surge Peak system pressure measured as a function of restricting or blocking fluid flow.

Suspended solids Solids that do not dissolve in liquid. Those that remain suspended and can be removed by filtration.

Suspension Any liquid containing undissolved solids.

Swing bolt Type of housing head closure that reduces service time. Opposite of thru-blot flange where studs are used, such as with ASA-type flanges.

T

Tare A deduction of weight, allowing for the weight of a container or medium; the initial weight of a filter.

Tensile strength Resistance to breaking. The amount of force required to break a membrane by stretching.

Tensiometer Device used to read the surface tension of a liquid or to read the interfacial tension between two immiscible liquids.

Terminal pressure Pressure drop across the unit when the maximum allowable pressure drop is reached.

Terminal velocity Steady velocity achieved by a falling particle when gravitational forces are balanced by viscous forces.

Three-stage filter separators Liquid prefilter coalescer separators containing three kinds or types of replaceable elements.

Throughput The amount of solution passing through a filter prior to plugging.

Tortuosity A continuous path that can be traced from a point on the upstream side of a filter to a point on the downstream side through a twisting pore pathway, traveled by the liquid or gas during filtration.

Tortuous path Crooked, twisting, or winding path that tends to trap or stop solid particles, commonly referenced in relationship to the flow pattern and makeup of a filter medium.

Total dissolved solids Portion of the total solids in the sample that passes through the filter and is indicated by the increase in weight in the vessel after the filtrate has been dried at 356°F.

Total solids/suspended solids The material residue left in the vessel after evaporation of a sample and drying in an oven at 217−221°F. The increase in weight over that of the empty vessel represents the total solids. Used in analyzing drinking water.

Total suspended solids (TSS) An indication of all the measurable solid matter in a sample of liquid.

Triboelectricity The charge of electricity that is generated by friction such as rubbing.

Triboelectric series (potential/charge) An inherent natural or induced positive or negative polarity charge that many materials possess. Fibers or a filtration medium with a triboelectric potential will capture charged and potentially neutral particles, assuming both positive and negative properties on the surface of the material. Triboelectric properties only work in air filtration assuming relative humidity below 90%.

True density Mass of a particle divided by its volume, pores, etc. being excluded from the volume calculation.

Turbidimeter An instrument for measurement of turbidity in which a standard suspension usually is used for reference.

Turbidity Any insoluble particle that imparts opacity to a liquid. A reference point to the total amount of solids contained in a liquid.

Turbulent flow Flow regime in which the flow characteristics are governed mainly by the inertia of the fluid. Turbulent flow in ducts is associated with high Reynolds number (Re). It also gives rise to high drag.

U

Ultrafiltration (UF) A separation method operating at 50−200 psi in cross-flow filtration mode. Efficiency is approximately 90%. Used to separate large molecules according to their molecular weight.

Uniformity of feed Uniformity of the mixture of the solids in the feed liquid.

Unloading The release of contaminate downstream that was initially captured by the filter medium.

Upstream side The feed side of the filter. Fluid that has not yet entered the filter.

Useful life Determined when contamination causes a filter or system to have an adverse (lower) flow rate, low efficiency, or high differential pressure, providing for an inefficient operation.

V

Vacuum Depression of pressure below atmospheric pressure.

Validation Demonstration that a process or product does what it is supposed to do by challenging the system and providing complete documentation.

van der Waals forces The relatively weak attractive forces that are operative between neutral atoms and molecules that arise because of the electric polarization induced in each of the particles by the presence of other particles.

Velocity head Velocity pressure or kinetic pressure.

Vent filters Filters that allow the passage of air while restricting the flow of fluid. Typically containing low micron rated microporous membrane media. Common in medical devices and pharmaceutical tanks.

Vessel A container, usually used as alternatively to the word housing, e.g., filter vessel.

Vibratory sifter Process equipment that separates solids by size on a metal screen through a vibrating action. Larger particles remain on the screen as fines fall through, sometimes to one or more higher mesh count screens for further separation of particle size.

Viscosity Degree of fluidity. Resistance to flow as a function of force or gradual yielding of force. For a given filter and differential pressure, flow rate will decrease as viscosity increases.

Viscosity index Numerical value assigned to a fluid indicating to what degree the fluid changes in viscosity with change in temperature.

Void volume The amount of open or empty area across the full spectrum of a material or substance. A term often used to describe the amount of porosity in a filter medium.

Volumetric flow rate Fluid flow expressed as a volume flowing per unit of time (cc^3/sec, ft^3/min, etc.).

W

Warp The yarns that run lengthwise or in the machine direction in woven goods.

Water flow/flux Measure of the amount of water that flows through a filter, a variable of time, the degree of contamination, differential pressure, total porosity, and filter area.

Weight of solids Measure of solid particulate matter contained in a fluid sample.

Weir (i) A diversion dam. (ii) A device that has a crest and some side containment of known geometric shape, such as a V, trapezoid, or rectangle and used to measure flow of a liquid.

Wet strength Strength of a medium when saturated with water.

Wetting agent A surfactant added to a filter medium to insure complete intrusion (wetting) by a high surface tension fluid such as water.

Wire cloth Woven fabric from metal wire used as a screen, surface filter, or media support. Often used in sifting, belting, hydraulic filtration, etc. Most common wire used is stainless steel.

Wound tubes Also referred to as string wound filters.

Y

Yoke End cap used to hold a cartridge in place.

Z

Zeta potential The potential across the diffuse layer of ions surrounding a charged colloidal particle.

SUGGESTED FURTHER READING ONLINE

For further reading, visit the websites of the following manufacturers (listed in alphabetical order):

3M: <http://www.3m.com/3M/en_US/country-us/>

3V Tech: <http://www.3v-tech.com/en>

Andritz: <http://www.andritz.com/group.htm>

BHS-Filtration Inc.: <http://www.bhs-filtration.com/>

Bokela: <http://www.bokela.de/en/company.html>

DrM: <http://www.drm.ch/>

FL Smidth: <http://www.flsmidth.com/>

G Bopp: <http://www.bopp.ch/>

Gore: <http://www.gore.com/en_xx/index.html>

Heinkel: <http://www.heinkel.de/en/Products/produktuebersicht.php>

Komline-Sanderson: <http://www.komline.com/>

Mott: <http://www.mottcorp.com/>

Outotec: <http://www.outotec.com/>

Pall: <http://www.pall.com/main/home.page>

Rosenmund: <http://www.rosenmund.com/>

Sparkler: <http://www.sparklerfilters.com/>

Steri Technologies: <http://www.steri.com/>

Westech: <http://www.westech-inc.com/en-usa>

REFERENCES

Armstrong, B., Brockbank, K., & Clayton, J. (October, 2014). Understand the effects of moisture on powder behavior. *Chemical Engineering Progress* 25–30.

Carpenter, C. (2013). Selection of a solid–liquid separation device. Seminar given to St. Louis AICHE section.

Chase, G. G., & Mayer, E. (2003). Selection guides for SLS equipment. *Filtration News, 21*(3), 18–22.

Darcy, H. (1836). *Les Fontaines Publiques de la ville de Dijon.* Paris: Victor Dalmont.

Harrison, R. G. (October, 2014). Bioseparation basics. *Chemical Engineering Progress* 36–42.

Hoffman, J. (July, 2013). Lifecycle costs for capital equipment in the CPI. *Chemical Engineering* 36–43.

Ives, K. J. (1975). *The scientific basis of filtration, proceedings of the NATO Advanced Study Institute.* Leyden: Noordhoff.

Jacob, K., & Collins, R. (2014). Solid–liquid separations: Fundamentals and applications. AIChE Webinar. New York: American Institute of Chemical Engineers.

Konnikova, M. (2013). *Mastermind: How to think like Sherlock Holmes.* New York, NY: Penguin Books.

Nicolau, I. (2003). Kuchen bildende filtration von suspensionen und filterberechnung. *CIT, 75*(9), 1206–1220.

Schubert, H., & Rippberger, S. (2003). *Handbuch der mechanischen verfahrenstechnik. Weinheim.* Verlag Wiley-VCH.

Sulpizio, T. E. (2013). Advances in filter and precoat filtration technology. Presentation at the American Filtration and Separations Society in Boston, MA.

Tichy, J. W. (2007). *Zum Einfluss des Filtermittels und der auftretenden Interferenzen zwischen Filterkuchen und Filtermittel in der kuchenfiltration (Row 3, no. 877).* Dusseldorf: Fortschrittsberichte VDI.

BIBLIOGRAPHY

Anlauf, H. (1986). *Dewatering of filter cakes in vacuum/pressure and pressure/vacuum filtration processes (Row 3, no. 114)*. Dusseldorf: VDI Fortschrittsreihe.

Anlauf, H. (1994). Standardfiltertest zur bestimmung des kuchen- und filtermediumwiderstandes bei der feststoffabtrennung aus suspensionen. *Filtrieren und Separieren*, 63−70.

Anon. (1997). VDI 2762: Filtrierbarkeit von Suspensionen, Verein Deutscher Ingenieure.

Avery, Q. D. (2002). Using a filter press to optimize your drying process. *Powder and Bulk Engineering*, *16*(4), 29−37.

Blackwood, T. (2013). Separation theory and practice: An introduction to centrifugal separation. Presentation at the St. Louis Engineers Club in St. Louis, Missouri.

Dickenson, C. (1997). *Filters and filtration handbook* (4th ed.). Oxford: Elsevier Advanced Technology.

Grimwood, C. (2006). Scaling-up filtering centrifuges. *Filtration*, *6*(2), 113−118.

Lepree, J. (2012). Say good-bye to old-school centrifuges. *Chemical Engineering*, 19−22.

Long, J. (2014). What every operator should know about filtration. *Water Environment & Technology*, 58−59.

Micrometrics (2011). The effect of particle shape on particle size measurement. *Application Note*, *158*, 1−3.

Neuman, R. G. (2010). Blowing considerations in filter presses. *International Filtration News*, 29 (5), 24−32.

Patniak, T. (2012). Solid−liquid separation: A guide to centrifuge selection. *Chemical Engineering Progress*, 45−50.

Perlmutter, B. A. (2014). Thinking like Sherlock Holmes for process filtration technology selection. In AICHE spring meeting, Paper 26A, New Orleans, LA.

Perlmutter, B. A., & Pierson, H. (2009). Selection of solid−liquid separation equipment: Part 1. *The Chemical Engineer*, 48−50.

Perlmutter, B. A., & Pierson, H. (2009). Selection of solid−liquid separation equipment: Part 2. *The Chemical Engineer*, 53−55.

Perlmutter, B. A., & Pierson, H. (2009). Selection of solid−liquid separation equipment: Part 3. *The Chemical Engineer*, 48−50.

Rushton, A., Ward, A. S., & Holdrich, R. G. (1996). *Solid−liquid filtration and separation technology* (1st ed.). Weilheim: VCH.

Schmidt, P. (2010). Filtration centrifuges: An overview. *Chemical Engineering*, *117*(13), 34−38.

Sentmanat, J. M. (2011). Clarifying liquid filtration. *Chemical Engineering*, *118*(10), 38−47.

Sentmanat, J. M. (2013). Engineering principles of precoating. *Filtration News*, 40−46.

Sentmanat, J. M. (2014). Filter problems and why filters fail. *International Filtration News*, 33 (3), 28−35.

Sparks, T. (2012). Solid–liquid filtration: Understanding filter presses and belt filters. *Filtration + Separation,* 20–24.

Stamatakis, K., & Tien, C. (1991). Cake formation and growth in cake filtration. *Chemical Engineering Science, 46*(8), 1917–1933.

Tichy, J. W. (2005). Ausgelegt und optimiert: Genaue filterversuche zur fest-flüssig-trennung. *CIT +, 8,* 62–63.

Tichy, J. W., Rippberger, S., & Esser, U. (2005). Einfluss des filtermittels auf den spezifischen Durchsatz von kontinuierlich betriebenen filtern. *Filtrieren und Separieren, 19,* 162–165.

Tiller, N. (1975). What a filter man should know about theory. *Filtration & Separation, 7/8,* 386–394.

Vastola, M. (2000). Testing: The logical way to select a centrifuge. *Powder and Bulk Engineering, 14,* 63–78.